TALES OF THE NORSE GODS

Oxford Myths and Legends in paperback

*

Tales of the
Norse Gods

Retold
by
BARBARA
LEONIE PICARD

Illustrated by
JOAN KIDDELL-MONROE

OXFORD UNIVERSITY PRESS
OXFORD   NEW YORK   TORONTO

Oxford University Press, Walton Street, Oxford OX2 6DP

Oxford New York Toronto
Delhi Bombay Calcutta Madras Karachi
Kuala Lumpur Singapore Hong Kong Tokyo
Nairobi Dar es Salaam Cape Town
Melbourne Auckland Madrid

and associated companies in
Berlin Ibadan

Oxford is a trade mark of Oxford University Press

© Barbara Leonie Picard 1953
First published in *Tales of the Norse Gods and Heroes* 1953
This selection first published in paperback 1994

A CIP catalogue record for this book is available
from the British Library

ISBN 0 19 274167 5

Printed in Great Britain

For
CARLIE

Preface

IN this book are some of the stories told by the Norsemen who lived in Scandinavia from about 2,000 to 1,000 years ago. We know them better, perhaps, as the Vikings, who sailed their ships along the coasts of Europe, plundering and laying waste, and, finally, settling and making their homes in France, in the Netherlands, in England, Ireland, and the Orkneys, and in bleak little Iceland. Even as far as Greenland they sailed, and to America, long before Columbus.

The people of the northlands were a nation of bold warriors and fine courageous women, who admired above all things strength in battle and bravery against great odds, and considered themselves disgraced for ever if they let a wrong to themselves or to their families go unavenged. They were a simple people, too, who enjoyed the simple things: good food and plenty of it, good ale to drink, and a crafty trick that could make them laugh.

Here are the stories of their gods, gods who were even such as every Norseman longed to be: brave, dauntless warriors or cunning tricksters, with their lovely, loyal wives, for ever striving against the hated giants who were the pitiless northern snows and frosts, and the grim northern mountains.

Some of the Norse names in the stories may seem a little difficult, both to spell and to pronounce; but at the end of the book there is a note on the pronunciation—it would be wise to read it before the stories—and also an alphabetical list of all names mentioned in the tales, with a note about each one. With the help of these, the difficulties should be much less.

Contents

Tales of the Norse Gods

TALES OF THE NORSE GODS

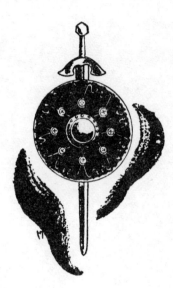

I

The Beginning of All Things

THIS is the story the Norsemen told of how the world began. In the very beginning of things there was only a vast chasm, Ginnungagap, with, to the north of it, Niflheim, the home of mist and darkness, and to the south, Muspellheim, the home of burning fire.

When it touched the cold that rose from the chasm, the damp mist from Niflheim turned to blocks of ice which fell with a terrible sound into Ginnungagap; and as the fiery sparks from Muspellheim fell upon these ice blocks they sent up steam which turned to hoar-frost as it rose into the cold air. And thus, with blocks of ice and with rime, the chasm was slowly filled.

And out of the hoar-frost two figures took shape; one was a giant man, Ymir, the first of the ice-giants and the father of all evil creatures, and the other was the cow, Audumla. Ymir drank the milk of the cow and grew strong and flourished; while for her food, Audumla licked the blocks of ice for the salt that was in them. As the ice was melted by her warm tongue and her breath, at the end of the first day the hair of a head appeared. At the end of the second day the head and shoulders could be seen; and when the third day was passed, Buri the divine stepped forth from the ice. He was good and beautiful, and he was the father of all kindly creatures. And to him was born, by the means of magic spells, Borr, his mighty son.

Meanwhile, as Ymir slept, from the sweat that trickled from his body there sprang up another giant man and a giant-maiden; and from the soles of his feet came forth six-headed Thrudgelmir.

And there was strife between the good and evil beings, between the giants and Buri and his son Borr; and though the giants increased in numbers, they could in no way prevail, and for a long time the victory went neither to good nor to evil. But one day Borr took to wife a friendly giant-maiden named Bestla, and a son was born to them called Odin, who was very powerful, and mightier even than his father and his brothers.

And in time Odin led his brothers against the giants and slew Ymir, so that all the giants were afraid at the loss of their leader and would have fled away, but that they were drowned in the blood which poured from his wounds. Only two of them escaped, Bergelmir, son of six-headed Thrudgelmir, and his wife; and they fled to the very edge of space, where they made themselves a home and called it Iotunheim, the land of the giant people. And from them were descended all the frost and storm and mountain giants that later troubled the world.

Ymir's body was then rolled into the chasm and from it Odin created the universe. First, in the very centre, he formed Midgard, the world of men, from the flesh of the dead giant, and fenced it all about with his bushy eyebrows for a protection. From his bones Odin made the hills and mountains, from his teeth the stones and rocks, and from his curly hair the grass and trees and shrubs. And all around Midgard, like a sea, flowed Ymir's blood, and beyond this sea lay Iotunheim. Over all, as the sky, Odin set Ymir's skull, resting it upon the shoulders of four strong dwarfs, North, South, East, and West, that it might for ever remain in its place; and between Midgard and the hollow vault that was the sky, Odin scattered Ymir's brains as the softly drifting clouds.

Then, so that there should be light in this world that he had made, both by day and by night, Odin took sparks from Muspellheim and flung them across the sky for stars, and two larger sparks he set in two golden chariots to be the sun and the moon. In the chariot of the sun he placed a beautiful maiden to rule over the light of the day, while to a lovely youth he gave power over the moon. But they were not allowed to remain shining peacefully in the sky, for as soon as the giants spied them from far Iotunheim, they sent two great wolves, Skoll and Hati, to devour them, so that there might be darkness in the world once more. And that is why the sun and the moon are never still in their places, but must continually move across the sky, for ever flying from the hungry wolves.

When their first task of creation was accomplished, Odin and the other gods looked at the world which they had made, and they saw that creatures like maggots had bred in Ymir's

3

flesh, and crawled about above and below the earth. Some of these creatures were good and white and shining, and these the gods made into the elves of light, giving them the sunshine and the flowers and growing things to care for, and making them a home in Alfheim, between the earth and the sky. The other creatures were dark and misshapen, and of them the gods made the elves of darkness, to live for ever below the ground, mining gold and precious stones; and they became cunning workmen, skilled in all manner of crafts. But no one of them might ever come upon the earth except by night, for one ray of sunlight falling across his dark features would turn him to stone.

On a broad plain above the earth lay Asgard, where the gods lived. Here was their council chamber, and the great field where they met; and here, too, each had his own home, fair and bright. Bifrost, the rainbow, was the bridge between Asgard and Midgard, the earth, and along this three-coloured way would the gods walk to and fro.

At the will of Odin there came into being a great ash-tree, Yggdrasill, whose topmost branches overhung his halls. This tree had three roots, one in Midgard, another in Niflheim, and the third in Asgard itself. At each root there flowed a spring; that in Midgard was known as Mimir's well, where wisdom and understanding were stored under the care of the giant Mimir; and that in dark Niflheim as Hvergelmir, and beside this pool of Hvergelmir lived a dragon which gnawed night and day at the root of the tree, seeking ever to destroy it. But the spring that flowed by the root that was in Asgard was called the fountain of Urd, and on it swam two holy swans. This fountain was guarded by the Norns, three sisters who knew the past, the present, and the future, and each day they sprinkled the leaves of Yggdrasill with pure water from their well.

On the very top of Yggdrasill, above Odin's halls, brooded a mighty eagle with a hawk perched upon its head; and along the wide branches walked and browsed four stags and the goat Heidrun, who gave mead instead of milk for the gods to drink. A squirrel scurried up and down the tree, carrying tales and words of unfriendliness between the eagle and the dragon which lived by Hvergelmir, until they grew to hate each other.

In Midgard the flowers blossomed and the earth grew green and lovely, but there were as yet no men to enjoy this pleasant home that had been made for them. And then one day, as Odin walked along the seashore with Hönir, the bright god, his brother, and with Loki who ruled the fire, they saw two trees standing straight and tall. And Odin touched them and gave them spirit and human life; from Hönir they received sense and movement; while Loki laid his hands on them and gave them warmth and beauty and red blood to flow in their veins. And there on the seashore stood Ask and Embla, the first man and woman to be created; and from them were descended all other men and women.

The gods of the Norsemen were the Aesir and the Vanir. The Vanir were the gods of nature: Niord, the god of the shore and the shallow summer sea; and his son and daughter, Frey and Freyia, Frey who ruled over the elves of light and Freyia the goddess of love and beauty; and Aegir, the lord of the deep and stormy seas, with Ran his wife, who caught sailors in her net and drowned them. Aegir and Ran were not truly of the kindly Vanir, for they were cruel and more akin to the giants, but like the Vanir, they ruled over nature and were on good terms with all the other gods.

The Aesir were the gods who cared for men; Odin the Allfather, king of all the gods, wise and just and understanding; and Frigg, his queen, who presided over human

5

marriages; Hönir, Odin's brother, the shining god, who lived among the Vanir; large, noisy Thor, the god of thunder, Odin's son, who always had a special corner in his heart for the peasants and the poor and the dispossessed; Tyr, the brave god of war; Balder and Höd, the twin sons of Odin and Frigg, Balder the god of daylight, who was the most beautiful of all the gods, and Höd who was blind and ruled over the hours of darkness; Hermod, Odin's messenger; and Heimdall, the divine watchman who kept guard over Bifrost, the bridge between Asgard and the world.

And lastly there was Loki, who was neither of the Aesir nor of the Vanir, nor yet of the giant race; crafty red-haired Loki, quick to laugh and quick to change his shape, the god of the fire that burns on the hearth, good and kindly when it wishes, but a merciless destroyer when it leaves its proper place. From the earliest days Odin and Loki had sworn an oath of brotherhood; and it was this which so often saved Loki in later times, when his cunning tricks so much displeased the other gods.

In Asgard Odin had three palaces; in one the gods met in council; and in another stood his throne, Hlidskialf, which served him as a watch-tower from where he might see all that passed not only in Asgard, but in Midgard and Iotunheim, and even in the depths of dark Niflheim, the home of mist, as well. Here would he sit with his two ravens perched upon his shoulders. Each day he sent these birds flying forth across the world and each evening they returned to tell him of the happenings of the day. At his feet would lie his two wolves who followed him like hounds wherever he went in Asgard, and at the feasting would eat the meat that was set before him; for the Allfather lived on mead alone, and no food passed his lips so long as he was among the gods, though when he travelled through Midgard, he lived like other men.

6

Odin's third palace was called Valhall and was set in the midst of a grove of trees whose leaves were gleaming gold. This palace had five hundred and forty doors, and its walls were made of glittering spears and its roof of golden shields. To this hall came all those warriors who had died in battle, when death had passed away from them as a dream, to feast and tell tales of their deeds as living men, and to test their fighting skill on one another with weapons and armour made of imperishable gold. For the Norsemen were great warriors, and they believed that when a battle raged, Odin would send out his warrior-maidens, the Valkyrs, to ride across the sky and fetch the slain to Valhall, where they would be with Odin himself, and feast upon the flesh of the boar Saehrimnir, which, though slaughtered and roasted each day, came back to life each night, and drink of the mead provided by the goat Heidrun. Thus every Norseman longed, when the time came, to die in battle; and his greatest fear was that he should suffer a straw-death, and die in bed, lying on his straw-stuffed mattress. For the spirits of all those who did not die fighting went down to dark Niflheim.

Odin ever sought knowledge and wisdom, that he might use them to the good of both gods and men; and one day he went to Mimir's well, the fount of wisdom and understanding, which flowed by that root of the ash-tree Yggdrasill which grew in Midgard, and asked the giant Mimir to let him drink of the magic waters.

Mimir looked long at Odin before he answered, and then he said, 'Even the gods must pay for knowledge.'

'And what is the price of wisdom?' asked Odin.

'Give me one of your eyes as a pledge,' said Mimir.

Unhesitatingly, Odin plucked out one of his eyes and gave it to Mimir, and Mimir let him drink from the well, and straightway Odin was filled with the knowledge of all things

7

past and present, and even into the future could he look. And though his new knowledge gave him joy, it brought sorrow to him also, for he could now tell not only what was past, but also the grief that was to come. Yet he returned to Asgard to use his knowledge to help the other gods and those men who sought his aid.

And Mimir dropped Odin's eye into his well, and it lay there evermore, shining below the water, a proof of Odin's love of wisdom and his goodwill towards mankind.

II

The Building of the Citadel

WHEN Asgard was but newly built, the gods wished for
a strong citadel outside their walls that might withstand the
attacks of the unfriendly giants, should any of them chance
to cross the rainbow bridge Bifrost and reach even to Asgard;
and while they were considering the best way it might be
done, a stone-mason from Iotunheim came by and offered
them his help. From his smiles they saw him, though a giant,
to be no enemy, and they listened to his proposals.

'I can build you,' said the giant, 'a fortress that shall be
proof against any of your enemies, though all the mountain-
giants and all the frost-giants fall upon it as one man. And
this shall be done in no longer than three years. What say
you, Aesir and Vanir, shall I build your citadel for you?'

9

'If indeed your boast is true,' said Odin, 'your work would be much welcomed here. Yet tell us first, what reward will you claim when your task is done?'

'Give to me the sun and the moon from the sky, that they may serve as lamps to light my house, and give me Freyia of the Vanir for my wife, and I will think myself well paid.'

'Well paid would you be indeed,' exclaimed Niord, king of the Vanir, angrily. 'My daughter weds with no giant.'

And all the gods who were present murmured against the stranger's words, that they were presumptuous and over-bearing.

But Odin held up his hand for silence. 'Your demands are great,' he said, 'I wonder will your achievements match them. But even should they, it were a shame to take from the sky the sun and the moon and leave the world of men in darkness.'

The giant shrugged his shoulders. 'That is the payment which I ask, that and no other. If the price is too high, then must you do without your citadel. I care not.'

Odin considered a moment and then he said, 'This is a matter which cannot be decided lightly, it must be debated by the council of the gods.'

So he called the gods to his hall, and they gathered together; all save Thor who, as was often his custom, journeyed in the north, seeking adventure; and they talked long about the giant and his offer. 'It is true that we need a mighty fortress, and if this giant's words are not false, then he can build us what we want, far better than we ourselves could fashion it,' they said. 'But the reward he asks is too great, we cannot give to him the sun and the moon, and Freyia is not for any giant to wed.' And at last they decided, with regrets, that they must refuse the stranger's offer.

Then Loki, god of the firelight, spoke. 'It seems to me,' he said, 'that we should be fools to let this workman go without making him serve us, as well he can if what he boasts is true. He has said that in three years his task will be done, and our citadel standing strong and mighty. Let us say to him that if it is completed in the length of but one winter, we will give him the reward that he asks. Then, if he accepts our conditions, so shall we have our fortress, and so shall we keep the sun and the moon and lovely Freyia, for no builder, however skilled, could finish such a task alone in the space of but one winter.'

'I do not doubt,' said Niord, 'that he will refuse.'

'We can but make the offer,' said Loki. And the other gods agreed with him, and together they went to the giant and told him of their conditions.

'If the fortress is completed,' they said, 'before the first day of summer, and if you have worked alone, without anyone to help you, the sun and the moon and Freyia are yours. What say you to our offer?'

The giant considered, and then he said, 'If you will let me have with me my horse, to help drag the blocks of stone, then will I accept your conditions.'

The gods hesitated, and Odin said, 'If the work is done with the help of a horse, then it is not done alone.'

But Loki laughed. 'What is a horse, brother Odin, but a horse? It has no hands to build with. Let him have his horse to help him carry the stones, he will have work enough with the building to keep him labouring well into the summer.'

So, persuaded by Loki, the gods agreed, and the giant went to fetch his horse. On the first day of winter he returned, leading the horse, a great black stallion which he called Svadilfari, and he set to work at once; and before long it was apparent that the horse was worth two men. Svadilfari had no hands to build with, but all night long he dragged blocks of stone for the giant, and on into the short winter's day; and while the daylight lasted, the giant piled stone upon stone into the semblance of a fortress. And so the work went on apace, while the gods and goddesses watched anxiously, and none more anxiously than Freyia.

And when there were but three nights to go until the first day of summer, the citadel was completed, all but the gate and the gateposts, and Odin called the gods to council, that they might think how best to save the sun and the moon and keep Freyia out of Iotunheim; and they could see no way by which to do it.

'Let him who advised our bargain now find a means to evade it,' said Niord bitterly. 'Loki is ever wont to give us bad advice.'

And they all turned to where Loki sat and demanded that he should think of a trick whereby they might be saved from giving the giant his reward, and threatening him should he fail.

'I will find out a way, never fear,' said Loki. But they were angry with him and all spoke against him, so he rose and went quietly from the hall.

He went to a wood close to where the fortress towered high, and took the shape of a dainty-stepping grey mare; and that evening, when the stone-mason came by with his weary stallion dragging a huge block of stone, Loki left the shadow of the wood and whinnied. Svadilfari looked up and saw the mare and whinnied in reply, while the giant sought to urge him on with his load of rock. But when Svadilfari saw the mare turn as if to trot back into the wood, he broke the traces and galloped after. Away went Loki through the wood like the wind, with Svadilfari after him; and the giant, calling in vain to his horse, following them, but a long way behind.

All that night the giant roamed in the wood, seeking

Svadilfari; but in the morning he had to return alone to his work, dragging the huge blocks of stone himself and setting them in place. And by the first day of summer the work was still not completed, and the gate was yet lacking.

The gods were glad when they saw how the sun and the moon might remain in the sky to give light to men, and how Freyia would not have to go into Iotunheim, and how they had a fine impregnable fortress as well, all but the gate, and that they could build for themselves.

When the giant saw the smiles of the gods, he first grew angry; and then he grew suspicious, remembering the grey mare which had come out from the wood where he had never seen a horse before; and then he grew angrier still, crying to the gods that they had cheated him.

'How have we cheated you?' they asked. 'Prove it to us, and we will pay you all we owe.'

But he could in no way prove his words, and they laughed at him, while he stormed and raged and threatened them so greatly that they ordered him forth from Asgard with all speed. But he would not go, and vowed to be revenged.

At that moment Thor returned from his journeyings, and hearing the giant's shouts, came at once to see what a giant did in the home of the gods. When he heard the threats, he too grew angry, and became very mighty in his wrath. 'Insolent giant,' he thundered in his great voice, 'you shall pay dearly for those words.' And raising high the weapon he carried, he struck down the stone-mason and made an end of him.

Thus did the gods win a stronghold against their enemies and pay no price for it.

But to Loki, as a mare, was born an eight-legged foal, with all the strength of Svadilfari, his sire, and double the swiftness, because of his eight legs; and with all the grace of the grey

13

mare that was the sly god of fire. And when Loki took again his own shape, he called the foal Sleipnir and gave him to Odin to be his horse; and of all horses brave Sleipnir was the best, and much beloved by the Allfather.

III

The Mead of Poetry

AMONG the Norsemen, the skalds, the poets who sang of
the deeds of the gods and the heroes at the feasting, and who
told tales of war and adventure during the long winter evenings when men were forced to stay at home, were held in high
honour and thought to be inspired by the gods themselves.
This is the story of the beginning of poetry.

When the world was very young, a dispute arose between
the Aesir and the Vanir and they all met together to settle
their differences, declaring that nevermore should there be
anything but peace between them. And as a pledge of their
everlasting friendship each one of them spat into a golden
vessel, and from their spittle they fashioned a man and named

him Kvasir. This Kvasir was so wise that there was no question that he could not answer, and he went about upon the earth giving freely of his knowledge to all who asked of it, and he was greatly beloved.

But there were two of the elves of darkness, the dwarfs Fialar and Galar, whose best delight it ever was to do harm to others of kindlier disposition than themselves, and they looked at Kvasir and watched him and hated him for his goodness. So one day when Kvasir came by the cavern where they lived, they called to him, saying they had a question to put to him, and bade him stop and enter their home.

'Willingly shall I talk with you, good dwarfs,' said Kvasir, and stepped from the kindly light of the sun into their underground dwelling.

But it was no question that Fialar and Galar wished to ask him, for the moment that he came among them, they struck him down and killed him, rejoicing at their wickedness.

'That was well done, brother, was it not?' laughed Galar.

'It was indeed well done,' replied Fialar. 'And now let us take the blood of Kvasir and mix it with honey and make mead of it, for I am certain that the blood of such a one as Kvasir was would give great knowledge and wisdom to him who drank it, and the possession of such mead might serve us well one day.'

So the dwarfs caught Kvasir's blood in two vats and a cauldron and mixed it with honey, so that they had mead which had the power of giving to him who drank of it not only wisdom and understanding, but the gift of words by which to pass on that wisdom and understanding to other men in songs which would make glad their lives. Yet the dwarfs did not drink one drop of the mead themselves, for they cared nothing for knowledge or poetry; instead they

16

hid in their dark cavern the three vessels which held it, against a time when they might find it useful.

And the gods, watching from Asgard, saw how Kvasir no longer went about the world, helping men with his wisdom, and they asked concerning him; yet no one could tell them where he might be. But only Fialar and Galar laughed and said, 'His words were so wise, we have no doubt that as they rose to his lips they turned in his throat and choked him, and he lies dead somewhere.'

And time passed and the two evil dwarfs looked around for someone else whom they might harm, and their eyes fell upon Gilling, a giant who lived with his wife on the shores of the sea. So late one evening, when it was growing dusk and they might venture out without being turned to stone, they dragged an old boat down to the beach where Gilling walked and greeted him.

'Good day to you, friend Gilling,' said Fialar. 'We go fishing. Will you not come with us and share our catch? Our boat is large enough for three, even though one be a giant.'

And Gilling was glad and went with them, calling out to his wife that he would soon be home.

But the boat was old and leaky, and under the giant's weight, she slowly filled with water. 'Should we not turn back and row for the shore?' asked Gilling anxiously.

'It is nothing,' laughed the dwarfs. 'Surely Gilling is not afraid to wet his feet?'

And Gilling said no more, for fear of their mockery. But soon it became apparent that the boat was going to sink; and with a shriek of laughter, Fialar and Galar jumped overboard and swam easily for the shore. But as they had known, Gilling could not swim, and the boat sank and he was drowned, and the two dwarfs were well pleased.

'Now let us go and make an end of Gilling's wife,' said Galar; and they ran off to Gilling's house.

Outside the house there was a millstone lying. 'Take you the millstone and climb on to the roof,' said Fialar, 'and when she comes out through the door, drop it upon her head.'

So Galar climbed on to the roof and held the millstone above the doorway, and Fialar knocked upon the door. 'Open, open, goodwife,' he called, 'for I bring you sad tidings.'

Gilling's wife opened the door and let him in, and Fialar told her that her husband had been drowned. He made no great show of sorrow when he told her, but she was too grieved to notice, and wept unceasingly.

'Come, goodwife,' said Fialar, 'it will soon be dawn, and if you will go down to the beach with me I will point out to you the place where our boat sank and your husband died. It may ease your heart to see his cold sea-grave.'

He went to the door and opened it, and she came with him. At the threshold he leapt nimbly back, and as she stepped out through the doorway into the grey half-light, Galar dropped the millstone on her head, and she died in an instant.

Fialar laughed delightedly as Galar jumped down from the roof to him. 'It is a good thing done, brother,' he said. 'I had grown tired of her weeping.' And together they ran back to their cavern before the sun rose and caught them.

But Gilling had two sons named Suttung and Baugi, and when Suttung heard how his father and mother had died, he strode over to the cavern and caught the two dwarfs as they came forth at dusk. He carried them, one in each hand, to the seashore, and flung them out on to a rock which was covered by the water at high tide. 'There may you stay and drown,' he said. 'Swimming will not help you now, for if you reach the shore I shall be here, waiting for you, and back you will go into the water, as often as you try to escape me.'

Fialar and Galar were very much afraid, but in vain they pleaded for their lives. 'You had no pity on my father and my mother,' said Suttung. 'Why should I spare you?' And he waited on the shore for them to try to save themselves by swimming to the beach.

When they saw that no pleading could move him, the dwarfs grew silent and thoughtful. Then suddenly Galar whispered, 'Brother, why do we not offer him the mead we made from Kvasir's blood? It is our greatest treasure, surely it was kept for such a time as this?'

'That is well thought of, brother,' said Fialar.

So the dwarfs offered their precious mead to Suttung in exchange for their lives, and after a hesitation, Suttung accepted the price, and allowed Fialar and Galar to swim to the shore. 'It will go ill with you if you try to cheat me,' he warned them.

But the dwarfs did not cheat him. Thankful to have escaped drowning, they gave him the three vessels which held the mead, and Suttung carried them off to his home. Yet like the dwarfs, the giant cared nothing for wisdom or poetry, so he drank none of the mead, but hid the two vats and the cauldron in the very heart of a mountain close by his house, and set his daughter, Gunnlod, to guard it night and day.

Yet there is nothing so secret that rumour of it will not travel forth, and in time the gods learnt of the mead that had been brewed from Kvasir's blood and how it was sealed away in a mountain cavern, guarded by a giant-maid; and the gods thought, 'This mead should be ours, to give to whom we please, as inspiration.'

And Odin said, 'I will go to Iotunheim and fetch this mead for our use.' And he went from Asgard alone.

With a grey cloak about him, such as travellers wore, a wide-brimmed hat pulled well down over his one eye, and with a staff in his hand, Odin set off for Suttung's house. On the way he passed a field which belonged to Baugi, Suttung's brother. In this field Baugi's nine thralls were mowing hay, and Odin stood on the edge of the field and watched them at their work. They toiled slowly, for their scythes were blunt, so taking from his belt a hone, Odin called out to them and asked if they would care to have him whet their scythes. Eagerly they brought the scythes to him and Odin sharpened them upon his hone. When the thralls set to work again, they found that their scythes were sharper than they had ever been before. 'That hone is better than any in our master's house,' they said to each other. 'Why should we not have it for ourselves?' So they called out to Odin, where he yet stood, watching them, and asked if he would sell his hone to them.

'Willingly,' said Odin, and he named a price, and they agreed to it. But they immediately began to argue amongst themselves as to who should keep the hone when once they had bought it, and they fell to quarrelling.

'Let him who catches it keep it,' called Odin, and he threw the hone among them.

With eager shouts they rushed at it, each striving to be the first to pick it up, thrusting one another aside and striking each other with their keen scythes, and, in their greed, wounding one another so grievously that in a little while there lay nine dead thralls in a field of unmown hay. And Odin took up his hone and went on his way.

At dusk he came to Baugi's house and begged a night's lodging and a meal. As he sat among the servants he heard how the master of the house grumbled at the loss of his nine thralls who had slain each other while the hay was yet uncut;

and he rose and went and stood before Baugi and spoke quietly. 'I will bring in your hay for you,' he said, 'and that as fast as your nine men.'

Baugi looked at him carefully, and something in the stranger's air of quiet confidence made him believe his words, idle boasting though they would have sounded on any other lips. 'What is your name?' he asked.

'I am called Bolverk,' replied Odin.

'What wage do you ask?' said Baugi.

'No more than a draught of the mead which the dwarfs gave to Suttung your brother, when the hay harvest is gathered in.'

'The mead was given to my brother,' said Baugi, 'I cannot promise you what belongs to another. Nevertheless, when your work is done and my harvest is stored, I will go with you to Suttung and ask that he lets you drink of his mead. Will you chance his reply?'

'I will chance it,' answered Odin.

So Odin stayed and worked for Baugi until the end of the season, and he worked as fast as nine men and better. And when the harvest was all gathered in, he went to Baugi and said, 'My work for you is finished, let us now go to your brother, that I may have the wage I have earned.'

They went together to Suttung's house, and Baugi told his brother of the bargain he had made with the stranger and asked that Suttung might help him keep it. But Suttung was angry and refused, blaming Baugi for promising a reward that was not his own to give, and upbraiding him; so that Baugi left his brother's house ill pleased at the words that had been spoken to him.

Odin saw his frowns and said quietly, 'Why do you not take what he has so churlishly refused you? Yonder stands the mountain where the mead is hidden. Why do we not go there and see how we may enter in and reach Suttung's treasure?'

And because he was angry with his brother, Baugi agreed, and they went together to the mountain; and there Odin took from his belt an auger and handed it to Baugi. 'Bore me a hole through the rock that I may enter the mountain,' he said.

Baugi took the auger and looked long at Odin, and a suspicion came into his mind. 'Are you not of the Aesir?' he asked.

Odin laughed. 'There is no one called Bolverk among the Aesir. Hurry, Baugi, and bore me a hole through the mountain to Gunnlod's cavern.'

So Baugi bored long with the auger through the hard rock of the mountain, and at length he drew the auger out. 'There is your hole,' he said.

But Odin blew into the hole and his breath sent the chips of stone flying back into his face. 'You have bored deeply,' he said, 'but you have not yet bored through the mountain.'

'There is no cheating you,' thought Baugi, and he shrugged his shoulders and bored yet deeper into the rock. And at last he had bored a hole right through the mountain even into the cavern where the mead was hidden. 'I have done as you asked, Bolverk,' he said. 'Now enter the mountain if you can.'

Swiftly Odin changed himself into a snake and slipped into the opening; and as swiftly, Baugi struck at him with the auger, but missed him, and was angry.

Odin wriggled through the hole that Baugi had made and came at last to the cavern where Gunnlod guarded the two vats and the cauldron filled with the mead, and he stood before her in his own shape, shining and splendid, the king of all the gods.

22

And the lonely giant-maid looked up and saw him standing there, and watched him long and unsmilingly. She was lovely, but her eyes were dark with bitterness and her lips had forgotten how to laugh. 'Surely you are of the Aesir?' she said at last.

'I am of the Aesir, fair one,' replied Odin.

Gunnlod rose and her voice trembled. 'I have waited long in this cavern all alone,' she said. 'My father has forgotten me and cares not what becomes of me so long as his treasure is safe. It matters not to him that I live unwed and childless, with no home that I may call my own, if only his mead is untouched. But the gods have remembered me at last and sent me a husband from among themselves. They have sent me a husband and much joy.'

Odin shook his head. 'Small joy will you get of a husband from the gods,' he said gently, 'for I can stay with you no longer than three days. Yet because of those three days you will never be forgotten, always will you be honoured in the minds of men as Gunnlod, Odin's giant-wife, and the mother of his son. But the choice shall be yours, fair Gunnlod. Let it be as you wish.'

She looked at him long and knew that she loved him, and at length she smiled, for the first time in many months. 'I will be Odin's wife,' she said.

He took her hand. 'As a pledge of our love,' he said, 'let me drink of your father's mead.'

'You may drink,' she said, 'but no more than one draught from each of the three vessels, for fear of my father's wrath.' And she let Odin drink from the first vat. Odin took one draught, but in that draught he emptied the vat. On the second day she let him drink from the second vat, and he took one draught, but in that draught he emptied the second vat. On the third day she let him drink from the

23

cauldron, and he took one draught, but in that draught he emptied the cauldron.

And on the morning of the fourth day, while Gunnlod still slept, Odin kissed her for the last time, without waking her, and turning himself once more into a snake, he wriggled through the hole that Baugi had bored in the rock. Once outside the mountain he changed himself into an eagle and flew back towards Asgard.

But Suttung had become suspicious after Baugi had asked him for a draught of mead as the price of Bolverk's hire, and from his house he had kept watch upon the mountain where the mead was hidden. So that when he saw a great eagle rise up and fly into the air, he guessed that a thief had been there, and he too took the shape of an eagle and set off in pursuit of Odin.

But the eagle that was Odin had a fair start and flew strongly, and reached Asgard safely. And when the gods saw it coming, they ran into the courtyard carrying a huge vat, and the eagle alighted on the rim of the vat and poured out the mead through its beak, so quickly that a few drops were spilled; and then took once more its rightful shape, and the gods welcomed back Odin amongst them again.

And from that day the gods kept the mead that was made from Kvasir's blood and gave it to those for whom they wished the gift of poetry, the true skalds and poets who gladden the lives of men. But the few little drops that Odin had spilled in his hasty flight from Suttung, these they left for any man to take; and that is why there will always be those who write bad verses and are a trial to their fellow men.

IV

Idunn and the Golden Apples

In the cavern in the mountain, where the mead of Suttung had been hidden, a child was born to Gunnlod, Odin's son Bragi, the god of poetry and eloquence, greatest of all the skalds. He quickly grew into a lovely youth and left his mountain home to travel all about the earth, spreading music far and wide.

One day as Bragi wandered, singing his sweet songs, he met with Idunn, the daughter of old Ivaldi, who was one of the dwarfs. Idunn was quite unlike her father and quite unlike her brothers, who were hideous and excelled in all smith's work, for she was fair and lovely to look upon and kindly in her nature, and differed from the elves of darkness in that she might look upon the sun without being turned to stone. Sometimes her father would allow her to leave her dark underground home and walk abroad by daylight, among the flowers and the green trees, and it was on such a time that Bragi saw her and fell in love with her. Happy to escape for ever from her gloomy home and her brothers, she went joyously with Bragi, and they set off for Asgard, where they were well received.

To Idunn was given the charge of the golden apples that the gods and goddesses ate to preserve their immortal youth. For being descended partly from the divine Borr and partly from the giant-woman Bestla, the gods had not of themselves everlasting youth or immortality, but won them both by eating these golden apples.

One day, when Odin, Hönir, and Loki were travelling through the world, they reached a lonely valley where a herd of fine oxen grazed, and being weary and hungry they killed an ox and flayed it, and lighting a fire, they spitted the ox over the flames and sat down to wait until it should be roasted. But though they waited a great while, the meat remained uncooked, however high they piled the fire with wood and however brightly it burned.

'There is some power here that wishes ill to us, that is plain,' said Odin.

'There is no one here but ourselves,' said Hönir. 'We are quite alone; how should the mischief of an enemy touch us here?' And he and Odin flung more wood upon the fire in a vain attempt to cook their meal.

But Loki rose and walked a little way off and looked about him all around that desolate place where only the oxen grazed peacefully, and after a time he saw, perched un-moving on a tree close by, watching the Aesir with malevolent eyes, a huge eagle. Loki returned to the others. 'We are not alone,' he said with a little smile. 'See, in yonder tree, where the big eagle watches us.'

'Surely a bird would wish us no harm,' said Odin.

Loki laughed. 'An eagle is not always an eagle,' he said. 'You, my brother, have taken such a shape yourself before this day.'

'That is true,' replied Odin.

Hönir called out to the eagle, 'Is it you, great bird, who will not let our ox-flesh cook?'

And with a screech the eagle answered, 'It is I, strangers.'

'Come,' asked Odin, 'why should you bear us ill will?'

The eagle did not answer the question; instead it said, 'If you will share your meal with me, then will your meat cook.'

'You are welcome,' said Odin. 'Come and join us, and may what you say be true, for we are hungry.'

The eagle flew down from the tree, and alighting by the fire, fanned the flames with its wings, and though they rose no higher than before, in a short time the meat was roasted and ready to be eaten. 'I caused the fire to burn,' screeched the eagle, 'so mine should be the first portion.' And it took more than half the meat for itself, and tearing it with its strong beak, devoured it at once.

The Aesir were angered when they saw their meal eaten by another. 'You have left but little for us,' said Hönir. Yet the eagle cared not for his words and snatched at the meat that remained and tore a great piece from it.

Loki picked up a stick from the heap of firewood and struck at the eagle with it. 'Begone, wretched creature that would eat all our food,' he cried; and with a last screech the eagle rose into the air. But Loki found that he could in no way lift off the stick from the eagle's body, and in no way take his hands from the stick, so that he was raised up, high into the air, and carried away by the eagle's flight, hanging from the end of the stick with his feet brushing through the tree tops. In vain he called to the eagle to stop, for it paid no heed to his cries.

'You evil monster,' said Loki, 'you are no bird. Who are you?'

'I am Thiazi, the storm-giant,' replied the eagle, 'and it has long been my wish to harm the gods.'

'Release me,' said Loki, 'and I will pay whatever price you ask.'

'Do you promise that?' asked Thiazi.

'I promise,' said Loki.

'Then swear that you will put into my power Idunn and her golden apples,' demanded Thiazi.

'I will give you all the goddesses, if only you will set me down,' cried Loki.

'It is only Idunn that I want. Idunn and the golden apples, so that I too, like the gods, may have immortal youth.'

'You shall have them,' promised Loki; and the next moment he was falling to the ground. He picked himself up and walked back to where Odin and Hönir waited, anxiously.

'Who was it?' asked Odin.

'What befell you?' asked Hönir.

Loki shrugged his shoulders. 'How am I to know who it was? But whoever it was, he soon became afraid of the rash game that he was playing and let me go.' And that was all that he would tell them. But though he said nothing of it, he thought deeply on the promise that he could not break, and planned how he might keep his word.

One morning, when Bragi was absent, Loki went to the fair groves of Brunnak where Idunn dwelt and greeted her kindly. 'What think you of this, Idunn?' he asked. 'In a certain wood, a little way beyond Asgard, there stands an apple-tree and it bears fruit that seems to me to be as marvellous as yours.'

'That could not be, Loki,' exclaimed Idunn. 'There are no other apples like mine.'

'But I have seen them growing on a tree. They are smooth and cool and golden, just as yours are, though whether they have the same magic powers, I cannot tell. Yet why should they not, since their appearance is the same?'

'I cannot believe you, Loki. Surely this is some jest of yours?'

Loki laughed. 'Had I dreamt that you would doubt me, fair Idunn, I would have picked you an apple and brought it to you, that you might see for yourself.'

'I wish indeed that you had,' she said, 'for I should like above all things to judge these apples for myself.'

'Then come with me,' said Loki, 'and I will take you to the tree, and then you may tell me if I am not right.'

Idunn rose. 'I will go with you, Loki.'

'Bring your own apples,' he suggested, 'that you may match them with these others.'

So Idunn went from Asgard with Loki, bearing her precious casket of magic apples; and he led her into a dark wood, and there Thiazi, like an eagle, flew down and snatched her up and bore her away to his home in Iotunheim; Thrymheim it was called, a bleak, wintry fortress high up in the mountains where the wolves howled and the winds swept through the pines until their branches groaned. Here Idunn was kept imprisoned, and every day Thiazi went to her and threatened or besought her that she might give him her golden apples to eat; but Idunn only clasped the casket to her and refused.

And the days went by and all the gods save Loki wondered where Idunn might be. 'She will be with Bragi,' they said at first, 'wandering through the world, leaving gladness wherever they pass. Soon she will return.'

But time went on, and Idunn did not return; and the gods thought how she had never left them for so long before, without a taste of the golden apples. And they grew tired, with eyes that had lost a little of their brightness, and faces that had become lined. And at last Bragi returned from his journeying, alone. He was in despair when he heard of the disappearance of his wife; and the despair of the other gods and of the goddesses was equal to his own, for now they had no idea where Idunn might be.

Odin called the Aesir and the Vanir together in a council, and one by one the haggard, ageing gods told when they had last seen Idunn, hoping thereby to find some clue that might lead to her recovery. And one of them told how he had last seen her, laughing and holding her casket of apples, going from Asgard with Loki.

'Stand forth, Loki, and answer this. Tell us if those words are true.'

Loki stood before the other gods, alone in the middle of the hall. There were streaks of grey showing in his dark-red hair and his face was thinner, but his eyes were as bright and as scornful as ever, and his smile as mocking.

'Where is Idunn, Loki?'

Loki laughed a little. 'She is in Thrymheim with Thiazi the storm-giant, and her apples are there with her.'

In an instant there was a consternation in the council of the gods, all crying out against Loki who had betrayed Idunn to the giants, and not only Idunn, but the youth and strength of all the gods as well; while Bragi wept despairingly for his lost wife.

At last Odin called for silence and turned to Loki. 'You deserve no mercy from us,' he said, 'and the worst fate we can devise shall surely be yours unless you fetch Idunn back to us, with her golden apples. How say you, will you undertake to go to Thrymheim, or will you face our wrath?'

And after a while Loki said, 'I will go to Thrymheim, and maybe I shall prevail and return with Idunn and the apples. Yet I do not go because I fear your threats, but only because I am becoming a little weary of growing old and ugly, I who was reckoned amongst the most beautiful of all the gods.'

So Loki took the shape of a hawk with keen eyes and strong wings, and he flew northwards to Iotunheim, to the big stone halls of Thrymheim, with a high stone wall around. And there he waited until he saw Thiazi come forth, and noted that he looked no younger than he should have been, and thought to himself, 'It seems that Idunn has not been kindly with her apples. It is well.'

Then he flew over the wall, and spying out with his sharp eyes the room where Idunn sat, pale and weeping, her casket of apples still clasped in her arms, he flew in through the narrow window, and seeing she was alone, took his own shape. 'Cease your weeping, Idunn. It is I, Loki, come to visit you.'

She looked up, and in her joy at seeing one of the gods again, she sprang to her feet and ran to him, and then stopped, remembering that it had been he who had betrayed her to Thiazi. She backed away from him and held her casket even more closely to her. 'What do you here, Loki? Have you come to mock at me?'

Loki laughed. 'I have come to carry you home again. We have all grown weary for a sight of your smiles, and, I must admit, for the taste of your apples. Will you come with me, or would you rather stay here with Thiazi?'

31

Idunn wept. 'Take me away with you, Loki. Take me back to Asgard.'

He spoke magic runes over her, and in a moment she began to shrivel, and the casket with her, until she was no more than a nut lying on the floor. Quickly Loki became once again a hawk, and picking up the nut in his claws, he flew out through the window, and southwards towards Asgard. But Thiazi saw the hawk flying from Thrymheim, and suspected that one of the gods had come to rescue Idunn, so he changed himself into an eagle, as before, on that day when the Aesir had met him, and flew in pursuit of the hawk.

And steadily the great eagle gained upon the little hawk, until Asgard came in sight and Loki felt that he could fly no farther. But still he flew on, clutching the nut in his talons; and the gods saw him coming and raised a shout from the battlements. And they ran and piled wood before the walls of their home and set it alight so that the flames rose high. And the hawk that was Loki dropped down into the blaze as the eagle swooped upon him. But Loki was the god of fire, and came unscathed through the flames, with Idunn safely in his claws; while the eagle's wings were burnt. The gods ran forward to slay Thiazi, and in an instant he lay dead, and Loki stood there in his own shape; and while the others crowded round, he spoke magic runes over the little nut in his hand, and it became Idunn with her casket, and with a cry she ran to Bragi's arms.

As soon as the first greetings were over, Idunn gave to each of the gods and goddesses one of her apples; even to Loki, who in the joy was quite forgiven for his treachery; and at once they grew young and fair and straight again, with bright eyes and smiling lips. And there was great rejoicing in Asgard.

V

Niord and Skadi

WHEN she heard how her father had been killed by the
gods in Asgard, Skadi, the daughter of Thiazi the storm-
giant, tied on her snow shoes, slung her quiver of sharp
arrows over her shoulder, and taking up her bow, set off for
Asgard to avenge his death. Right into the hall of the gods
she came and challenged all who would to fight with her.

The gods watched her standing there, brave and angry,
in her cloak of fur and her flashing golden helmet that was
yet no brighter than her hair that hung down below it, one
giant-maiden against all the might of the gods, and their
hearts warmed towards her in admiration of her courage.

35

'What good think you it will do, Skadi, if you too die in Asgard as your father died?' asked Odin gently.

'My father must be avenged,' said Skadi. 'Or are all the gods too much afraid to fight against one giant-woman?'

'Thiazi was our enemy,' said Odin, 'but there is no one of us here who would willingly shed your blood. Go in peace, good Skadi.'

'My father must be avenged,' she repeated, 'and there is but I to avenge him. Since I heard that he was dead I have not smiled, my heart is cold and dulled with grief, and I think that I shall never laugh again.'

'We would call you our friend, Skadi, and not our enemy,' said Odin. 'We would see you living and happy, not dead at the gates of Asgard. Come, accept atonement for your father's death and be reconciled with us. Let us give you in recompense the best that we have to offer. Choose for yourself, Skadi, a husband from among the gods, and live in peace with him, no longer the daughter of the storm-giant, but one among the goddesses.'

Yet all Skadi would say was, 'My father must be avenged.'

But the gods sought to persuade her, saying, 'We should welcome you with honour, Skadi, and rejoice to have you as our friend.'

'Thiazi's eyes shall I take,' said Odin, 'and cast them up into the sky, two new stars in the heavens, in memory of him, to show that though our enemy, we ever held him in respect.'

A little appeased by his words, Skadi looked around her at all the gods assembled in the hall; wise and kindly Odin, great tawny-bearded Thor, smiling gentle Niord, handsome red-haired Loki who had been the cause of her father's death, Balder the young god of sunlight, the most beautiful

36

of them all, with his fair white skin and his golden hair, and all the other gods before her. She looked at them all closely, thinking upon Odin's words, but ever her eyes returned to Balder, and at last she said, 'If I may have Balder for my husband, then I will forgive the wrong you did my father, and live in peace with you.'

'You may choose your husband for yourself, Skadi,' smiled Odin, 'but choose him by his feet alone, and abide by your choice.'

'There is no doubt but that Balder, being the most beautiful, will have the loveliest feet,' thought Skadi, 'and thus shall I know him instantly.' So aloud she said, 'To that I agree, but further demand that before I am to be your friend, you shall make me forget my grief, and laugh.'

'That will be a hard task,' said Odin, watching her pale unsmiling face.

'Not for me,' said Loki. And he rose from his place and going to Skadi set himself to make her laugh, with all the skill he had in jesting. And at his tricks and the tales he told, all the gods held their sides with laughing; but Skadi never even smiled, for she hated Loki who had caused her father's death, and she would have found it easier to laugh at the jests of any other of the gods.

But Loki's wit and cunning prevailed at last, and Skadi forgot her grief and anger and smiled a little, and then suddenly she laughed aloud.

'See,' said Loki, 'did I not promise that I should make you laugh?'

'No one could fail to laugh at your antics, when once you have set yourself to make him laugh,' said Skadi.

'And are we friends at last, fair Skadi?' asked Loki.

'We are friends,' she said. But in her heart she never forgave him for Thiazi's death, though his had not been the

37

hand that slew the giant, and though she lived in peace and friendship with the gods from that day.

'Now must you choose a husband for yourself from among the gods,' said Odin. And Skadi's eyes were covered so that she could see no more than the ground before her, and Frigg led her down the length of the hall, past the gods as they sat on the bench along the wall, and Skadi carefully watched the feet of each as she passed him by.

Of some feet she found it easy to guess the owner; there were Odin's, hardened by his countless journeyings among men; and the huge feet below strong ankles that could belong to no one else than Thor; but she passed by them all, until she came to a pair of feet so white that the veins showed blue through the skin, with delicately arched insteps and shapely ankles, and she smiled to herself. 'These will be the feet of Balder,' she thought. But she said nothing until she had seen the feet of all the gods and found no others so beautiful. Then she returned to where he sat who owned the feet she thought to be the feet of Balder, and said, 'Surely you must be Balder, the god of the sun, and I will have you for my husband.' And she took away the cloth that covered her eyes and found that she stood before the kindly, smiling Niord, king of the Vanir; Niord who was lord of the winds and the waves as they broke on the shore, whose feet had never been hardened by the rocky mountain ways trodden by the other gods, for he walked but rarely save across the sandy beaches, among the little pools left by the tide, and his feet were fairer and whiter even than those of Balder who was otherwise the most beautiful of all the gods.

Skadi was disappointed at her choice, that she had been mistaken. But Niord was handsome and kindly, a husband such as would have pleased any maiden, and she resolved to be content, and smiled at him when he took her hand and pledged himself to her before all Asgard.

Niord had a palace on the seashore, Noatun, where the sea birds cried and the young seals gambolled and the wild swans which were sacred to him gathered. From here he ruled the sea around the coast, calming the waves called up by Aegir of the deeper ocean, and protecting the fishermen from cruel Ran and her nets. For him the sunlight rippled on the water in the little bays and creeks, and the wind played gently in and out his halls.

To this palace Niord brought Skadi, showing her all its delights, thinking that it would please her. And indeed, at first it seemed pleasant enough in contrast to her own bleak home of Thrymheim, but before many days had passed, Skadi hated it. She went to Niord and said, 'For many days I have not slept for the noise of the sea at night, and the crying of the sea birds by day fills me with melancholy. The winds that blow here are too gentle, so unlike the mountain tempests, and the little rippling waves seem puny and childish beside the memory of my own mountain torrents. Good Niord, I fear I cannot live with you much longer here, or, sleepless and homesick, I shall go out of my mind.'

'I am grieved that my home displeases you,' said Niord. 'You must not suffer on account of me.' He thought in silence for a moment, then went on, 'Let us spend a part of our time together here, and a part of our time at Thrymheim. In every twelve nights, let us pass nine at your own home and three here at Noatun. How will that please you, my Skadi?'

Overjoyed, Skadi hastened to make ready to return to Thrymheim, and Niord went with her, away to the bleak mountain fastness in Iotunheim, far, far to the north, where the magpies nested, that were Skadi's own birds. Nine days out of every twelve they spent in the stone home that had once been Thiazi's, while the wind raged down the mountains through the pines, and the snow lay close around. And Skadi stood in the courtyard with three magpies on her shoulder, watching the clouds hurtling across the leaden sky and laughing with delight at the wildness of it all; while Niord, wrapped in fur cloaks, huddled by the fire and thought of his pleasant home, so far away.

And at last Niord said to Skadi, 'I cannot sleep in Thrymheim for the howling of the wolves by night, and each day is colder than the last, and here there is no joy or comfort. Nine nights out of every twelve are too many for me to remain in this place.'

'And three nights out of every twelve are too many for me to pass in Noatun, where the sea birds scream, and three days out of every twelve are too many, while the sun shines dazzlingly on the water and blinds my eyes.'

'What then shall we do, my Skadi?'

Skadi shook her head, 'I know not, only that it seems we cannot live together happily.'

They talked long and made many proposals, only to reject them all; and at last it seemed to them that Skadi had been right, and they could in no way live together. So Niord and Skadi bade each other good-bye, and Niord went back to live in Noatun, where the wind played gently with the sunlit waves, and his wild swans flew about him when he walked abroad, beating the warm air with their white wings; and Skadi remained in Thrymheim with her chattering magpies that perched on the beams of her halls, where the wind howled as loudly by night as the wolves, while she listened to it in the darkness with fierce joy.

And thus each of them was happy once again, meeting only now and then, in Asgard, on days of festival.

VI

Frey and Gerd

FREY, the son of Niord, ruled over the elves of light in Alfheim, and sent gentle showers and warm sunshine to make the earth fruitful. His were the fields in harvest time which brought riches to men; and his the flowers which decked the land when the cold frost-giants of winter had been driven away by spring. He was young, and fair to look upon, and happy.

But one day he dared to sit on Hlidskialf, Odin's throne which stood in Odin's watch-tower, where none but the Allfather himself might sit. Frey sat upon the throne, looking all about him, across Asgard and Midgard and even into

41

Iotunheim, and fancying himself, in jest, to be the ruler of all things known; and he laughed with pleasure at his game.

But as he gazed out over Iotunheim, his eye lighted on the house of Gymir, one of the fierce mountain-giants; and while he watched, the door of the house opened, and Gerd, Gymir's daughter, came forth. She was the loveliest of all the giant-maidens, and the moment that he saw her, Frey knew that he could have no other bride. But he knew, too, that no maid of Gymir's line would consent to wed with one of the Vanir, their hated enemies.

He watched her as she crossed the courtyard from the house to her bower, followed by her serving-woman, carrying a distaff and white wool. And when she had gone from his sight and he could see her no more, still the memory of her as he had seen her remained in his mind to trouble him, and it was with a heavy heart that he left the watch-tower and returned to Alfheim, his own home.

'Fitly am I rewarded,' he thought bitterly, 'for my presumption in mounting Odin's throne, for now shall I have no peace or joy until Gerd is mine. And that can never be.'

Frey spoke not of his sorrow or his love to any of the other gods; but he could not hide his grievous longing, and he went among his companions with a mourning countenance, and cared no more for feasting and revelry. Nor did it concern him how things went on earth, whether the crops were good or poor, or whether the flowers blossomed or no.

Niord watched his son, how sick at heart he seemed, and he was sad that Frey spoke not to him of his secret sorrow. At last he sent for Skirnir, who was Frey's loyal servant and his friend.

'Skirnir,' said Niord, 'for days past have I watched my son troubled by some sorrow of which he will not speak. I would that you could find what ails him, that it might be remedied. Go now and see if you can move him to tell you what is amiss, for he loves you well, and may speak to you.'

Skirnir went to Frey where he sat in his wide hall alone, with downcast eyes, and spoke with him. 'Lord of the elves,' he said, 'giver of good things to the world, will you not tell me why it is that you sit thus alone and sorrowing? For perchance there might be some little thing that I, even I, your servant, could do to ease your heart.'

Frey looked up at him and sighed. 'Even by you, good Skirnir, could I not be comforted.'

'At least tell me why it is you grieve,' pleaded Skirnir, 'for you have ever trusted me.'

And after a while Frey answered him. 'In Iotunheim, in Gymir's house, there dwells a giant-maiden, Gerd, Gymir's daughter. I did but see her once as she stepped forth from her father's door, but it seemed to me in that one instant as though the sea itself and the very sky shone more brightly in the reflection of her beauty, so fair she is. But I am Frey of the Vanir, and she is Gymir's daughter, and he and his kin hate us for ever, so Gerd may never be mine.'

But Skirnir was brave and loved his master, and he determined that Frey should win Gerd and no more sit disconsolate. 'Lord of the elves,' he said, 'I will go to Iotunheim, to Gymir's house, and speak with Gerd and win her love for you. Give me for the journey Blodughofi, your horse that goes through fire and water and darkness and never falters, and your sword that slays whenever it strikes, and I will return with the maiden or with her promise, or I will never return again.'

Frey's face lightened a little with hope, and on earth a few flowers raised their drooping heads and a few green ears of corn grew ripe. 'If you will do this thing for me, good Skirnir, not only for the journey will I give to you my

43

sword, but for ever, a gift from Frey to his best-beloved friend.'

He gave Skirnir eleven golden apples to carry with him, and a golden ring, tokens of his love for Gerd; but Skirnir, fearing lest gifts might not move her, secretly took a slip of wood and carved it with magic runes, staining them with a scarlet dye.

Then Skirnir girded on his master's sword and saddled and bridled his master's horse. 'Come, brave Blodughofi,' he said, 'we must go a hard road together, and together we shall come home at length, or in Iotunheim we both shall perish.'

Skirnir rode into Iotunheim, and a long and dreary way it was; but at last the house of Gymir came in sight, tall and wide, with a high wooden wall around it, and snarling hounds at the gate.

A herdsman sat upon a little hillock close by, keeping Gymir's flocks, and Skirnir greeted him and asked, 'Tell me, herdsman who can see so far from the top of your hill, how I may enter Gymir's house, for I would have speech with Gerd.'

The herdsman laughed. 'Gymir's daughter is not for such as you to speak with, stranger. You are rash, and indeed, I think that you are out of your mind, that you would enter Gymir's house. Ride back again the way that you came, or soon you will be dead.'

'I shall perish when the time comes for me to perish,' said Skirnir, 'and not before. And if it is today that I must die, well then, it is today.' And he set Blodughofi at the wall of Gymir's house, and the good steed galloped forward and leapt right over the wooden paling, while the hounds howled and snapped below.

In the house Gerd sat spinning with her serving-woman, and there came a great noise from the courtyard without, the hounds baying and a clattering of hoofs and a voice calling her. 'Go to the window,' said Gerd, 'and tell me what these sounds may mean, for they make the very walls of my father's house tremble and shake.'

The woman went to the window-place and drew aside the leather curtain that hung over it, and peeped out. 'In the courtyard a fair young stranger dismounts from a wondrous steed. He is like to no one that I have ever seen before in Iotunheim, and how he entered here I cannot tell.'

'Open the door,' said Gerd, 'bid him welcome and fill him a horn of mead, though I fear that one who comes so mightily can bode no good to us. I would that my father were home.'

So Skirnir entered Gymir's house and saw the lovely Gerd, and understood why Frey sat sorrowing all day. And Gerd laid down her distaff and looked at Skirnir closely, and questioned him. 'You are not one of the giant-people, stranger, and you are not one of the dwarfs, yet it is a mighty feat that you have performed today in entering my father's house. Are you of the elves or of the Vanir, or are you from Asgard, that you show no fear of Gymir's wrath?'

'I am not of the elves or the Vanir, nor of the Aesir, fair Gerd, yet have I come from Alfheim with love-gifts from great Frey. Here are eleven golden apples that he sends you, that you may listen to him kindly.' And Skirnir held out the apples to her.

But Gerd would not take them, and looked angrily at him. 'I accept no gifts from one of the hated Vanir. No tokens or words of love could ever pass between us. Go, tell that to your Frey.' And she picked up her distaff and went on with her spinning.

Skirnir held out the golden ring. 'Lest the apples might not please you, fair Gerd, my master sent you this ring, belonging to the gods.'

Gerd glanced at it scornfully. 'You may take it back to your master. What is gold to Gymir's daughter? My father has store of it enough. Farewell.' And she turned again to her spinning.

Then Skirnir drew Frey's bright sword, that slew whenever it struck. 'Though I am loath, fair Gerd, to harm one whom my master loves, yet shall I strike your beautiful head from your shoulders, if you will not be kind to him.'

The serving-woman cowered in a corner of the hall; but Gerd looked at him unafraid, her eyes hard and cold. 'I do not fear your sword, servant of Frey, nor yet your threatening words. But were my father here you would not speak so boldly, indeed, you would not speak at all, for he would have silenced your talking for ever.'

Skirnir sheathed the sword and took from his belt the slip of wood on which he had carved the magic runes. 'On a high hill,' he said, 'before the gates of Niflheim, sits the eagle who fans the winds with his wings. On that selfsame hill, wretched and alone, bound and tormented and mocked at of all, shall you pass your time in misery, if you refuse the true and faithful love of Frey. Your days shall be more anguish-laden than any you could dream of, food shall be hateful to you for that it prolongs your life, Odin the All-father shall be wroth with you for evermore, and Frey shall be your enemy, if his love you reject. Never a husband or a wooer shall you find, but dwell for ever lonely and hated of all, cursed by the gods and scorned by the giants, if for Frey you have no kindness. Magic runes have I written here to curse you with, fair Gerd. Very powerful they are, and this doom shall be yours unless you become Frey's bride.'

Gerd hid her face with her hands and shuddered, for she knew how mighty were the spells of the gods; and Skirnir watched her, waiting for her to speak; while the serving-

46

woman wailed in the corner of the hall, her head covered by her skirt.

At last Gerd looked up. 'Though it would be a shame to me, and worse than death, to wed with one of the hated Vanir, yet do I fear your terrible runes, so that I must obey you.' She sighed. 'Drink one last horn of mead in my father's house, and then return to Frey and tell him that he has won Gymir's daughter for a wife, though heavy of heart she is, and consents not willingly.'

'Before I go,' said Skirnir, 'you must say where you will meet him, and when he may await you, and give me your word to be there.'

'There is a wide and silent forest, green and leafy, named Barrey, which I do not doubt is known to both of us. There will I meet Frey, nine nights from now.' And Gerd wept.

But Skirnir laughed for joy that he had served his master well, and mounting Blodughofi, he rode away from Iotunheim.

Frey was standing before the gates of his house, awaiting him. He ran forward and caught at Blodughofi's bridle. 'Tell me, good Skirnir, had you any success?'

'In the forest of Barrey, nine nights from now, will Gerd await you, to be your bride.'

And Frey embraced his friend and said, 'The sword is yours, Skirnir, for never has anyone served me better. But, oh, how long will nine nights seem, until the time I can meet my Gerd!'

And when the ninth night was come, in sorrow Gerd went secretly from her father's house to the forest of Barrey, and there, with dread, she awaited the coming of Frey. 'Surely,' she thought, 'nowhere is there one more wretched than I, a giant-maid forced to wed with one of the hated Vanir.'

47

But then she looked up and saw Frey coming towards her, young and fair and bright, with sunlight in his golden hair even in the darkness; and she saw that he came as a lover, gentle and kind, and not as a conqueror, and she was unhappy no more.

And long they dwelt together in joy and peace.

VII

The Lady of the Vanir

BESIDES his son, Frey, Niord, king of the Vanir, had a daughter named Freyia, who, like her brother, was gay and happy and fond of the sunlight and the flowers. Freyia, the lady of the Vanir, was the loveliest of all the goddesses, and her home, Sessrumnir, was built on the wide space called Folkvangar. She was the goddess of love and beauty, and lovers prayed to her for joy and happiness and for the safety of their loved ones. When Freyia went abroad, she drove in a chariot drawn by cats, which were her favourite animals, while swallows flew about her, and cuckoos called their spring-like notes to welcome her, so that these two birds came to be held sacred in her honour.

Freyia was married to the god Od, and in his company she

49

found great joy, and would have been ever happy and contented, had he been always with her. But Od was a great wanderer, and it was his pleasure to leave Asgard often and travel everywhere, seeing all there was to be seen, often for many days on end. And in his absence Freyia sat alone in Sessrumnir, bored and discontented, and sometimes weeping. Her tears were drops of red gold, not brine like mortal tears, and they enriched the ground they touched. Sometimes Freyia would leave her home and wander through the world, seeking for her husband; and often it was hard to find him, so that she shed tears of loneliness and disappointment in many places; and that is why gold is dug from the earth in many different lands. Yet when she found Od, then there were no more tears, but only smiles and laughter, as hand in hand they came home to Asgard together, the swallows twittering about their heads, and the flowers springing up where Freyia's feet had trodden.

One day when Od was away from home, Freyia, seeking some consolation for his absence, wandered alone from Sessrumnir through the fields, picking summer flowers for her hair. Below a hill she came to a certain passageway which led down beneath the earth to Svartalfheim, where the dwarfs, the elves of darkness, lived and plied their trade as smiths, mining the metals and precious stones from which they made weapons and armour and jewellery surpassed by none.

Curious, Freyia peeped into the passageway, staring into the gloom, but nothing could she see; yet there came faintly to her ears the sound of a distant hammering, the clink of metal upon metal as the dwarfs went about their work. Since Freyia knew how fine was the craftsmanship of the dwarfs, and since she had nothing better to do in Od's absence, she thought to herself, 'I will go into Svartalfheim and see the dwarfs at work, it will while away the hours for me.' And she slipped into the dark passageway, which grew the lighter for her presence.

The farther Freyia went along the passage, the louder grew the sounds of hammering, while in the distance there appeared a glowing as of the furnace of a forge. Freyia walked towards this light, but on the way she passed an open doorway leading into a cave, and looking in, she saw four dwarfs sitting on the floor together, their heads bent close over some object which they were admiring.

'It is the finest thing that we have ever made,' said one of them.

'It is finer than any of our kinsmen could have wrought,' said another, 'and we do well to be proud of our skill.'

'It is fit for a goddess,' said a third dwarf.

'Indeed,' said the fourth, 'I do not doubt that any of the goddesses would be pleased to own it.'

Overcome by curiosity, Freyia slipped softly into the cave, trying to see, through the dim light, what it was that the four black dwarfs were looking at. Then one of them said, 'See, my brothers, is it not perfect?' And he held up above his head a necklace made of gold and gems, so bright that it lighted up the gloom of the cave and so beautiful that Freyia cried out in wonder at it. The four dwarfs heard her and leapt to their feet, turning around; and the one who held the necklace hid it behind his back. Suspiciously they stared at her.

'Who are you, stranger?' asked the first dwarf.

But before she had time to answer, the second exclaimed, 'Why, it is the lady of the Vanir, lovely Freyia herself, come to visit us. Are you not Freyia, fair one?'

'I am,' she said, never taking her eyes from the dwarf who held the necklace.

'We are honoured, brothers,' jeered the third dwarf, 'Freyia has paid us a visit. Hurry, fetch a stool that she may sit down.'

Two of them carried forward a wooden stool and placed it in the centre of the cave. The third dwarf made her a mocking bow, bending until his head touched the rocky floor. 'Be seated, fair guest,' he said.

But Freyia was still staring at the fourth dwarf, he who held the necklace hidden behind his back, and she noticed nothing else. 'What is that which you are holding?' she asked at last.

'A necklace we have made,' replied the fourth dwarf sullenly.

'Let me see it,' said Freyia.

'No.'

Freyia smiled and her voice was honey-sweet. 'Let me look at it,' she pleaded.

The fourth dwarf hesitated. 'What do you say, my brothers, shall she look at it?'

'What harm is there in looking?' laughed the third dwarf. 'By all means let her see it.'

The fourth dwarf held up the necklace and once again it lighted the cave, flashing like golden fire set with lesser lights of every hue. Freyia gazed at it, entranced.

'Look well on it, fair Freyia,' mocked the third dwarf, 'for it is not yours, and in Svartalfheim we keep what is our own.'

But Freyia never heard him; all else was forgotten as she stared in wonder on the necklace, a far lovelier jewel than any she possessed. At last she whispered, 'It is beautiful. Will you give it to me?'

Three of the dwarfs cried out indignantly, and the one who held the necklace hid it once again behind his back.

But the mocker held his sides with laughing and rocked upon his feet with mirth. 'Freyia wants our necklace,' he gasped. 'Freyia wants our necklace.'

'Give it to me,' begged Freyia. 'Good dwarfs, give your necklace to me.'

The dwarf who held it answered her. 'It is Brisinga-men, and we have made it with our own hands and laboured long at it. Brisinga-men is not to be given away to anyone who asks.'

'I will give you fair exchange for it,' said Freyia.

'What fair exchange could there be for Brisinga-men?'

'I will give to you all my other jewels, or the flowers that bloom where I have trodden, or the light that shines about my hair, or the spring-song of the cuckoo. Any of these or all of them, will I give you in payment for Brisinga-men.'

'What do we want with your jewels, we who have made Brisinga-men? What do we care for flowers, or the song of the cuckoo? And we have gold enough, and gems, in the heart of the earth, to shine for us, we do not need the brightness of your hair.' And all four of them laughed at her.

'Then tell me what price you will take for your necklace,' Freyia pleaded.

'It has no price,' said he who held it. 'It shall be ours to keep.'

'It is not for the Vanir to touch or to hold,' said another of the dwarfs.

But the mocking dwarf said nothing, and only watched her, his eyes alight with malice.

'I will pay any price you ask,' said Freyia.

'Brisinga-men has no price,' repeated the fourth dwarf sullenly.

But the dwarf who had mocked her whispered to the others so that she could not hear what he was saying, and

53

at his words they all stared at him, surprised, and then burst out into evil laughter, jumping up and down in their glee, so that Freyia once again caught sight of the necklace as the dwarf who held it stamped about in a dance of malignant delight.

'What price do you ask?' said Freyia again, fired by this latest glimpse of Brisinga-men.

The dwarfs turned to her and spoke together. 'She to whom we give Brisinga-men must willingly allow us to embrace her, that we may boast all through Svartalfheim that we have held the lady of the Vanir in our arms and kissed her.'

Freyia drew herself up, the necklace forgotten in her indignation, and her eyes flashed in anger. 'Black crawling vermin nourished by Ymir's rotting flesh, how dare you speak thus to a goddess?' And she turned and made to go from their cave.

But he who had mocked her called after her, 'Do not go without bidding good-bye to Brisinga-men, Lady Freyia. Look, see where Brisinga-men shines out in farewell for you.'

And Freyia turned and saw how all the dwarfs were holding up the necklace, high above their heads, their faces burning with mockery and malice, and their misshapen bodies shaking with mirth; and she paused and stared at the necklace, and hesitated. They shook it slowly from side to side, so that it glinted and sparkled. 'Look upon it well, fair Freyia, for you will never see it more.'

And Freyia looked, and in that moment she knew that there was nothing that she would not do to gain it, and goddess though she was, nothing which she would not suffer, that she might call it hers. 'I will pay your price,' she said.

They screeched with laughter, dancing about on the rocky floor of the cave and congratulating each other.

'Freyia will pay our price,' they screamed. 'The lady of the Vanir will buy our necklace.'

And Freyia waited, thinking only of the moment when she might snatch the necklace from their hands and run with it along the passageway, back to the daylight; while they ceased their mocking laughter and set to quarrelling among themselves as to which of them should kiss her first.

Brisinga-men remained Freyia's favourite jewel, and she wore it day and night, not even taking it off when she lay down to sleep. It was because he knew how much she valued it, that Loki made up his mind to steal the necklace in order to torment her, and he planned to go to Sessrumnir one night when Od was away on his wanderings and take it from around Freyia's neck while she slept.

Heimdall, the white one of the Aesir, the watchman of the gods, had a house beside Bifrost, the rainbow bridge that linked Asgard to the earth, and here he kept guard unceasingly to warn the gods of the approach of any of their giant enemies. Clad in white armour, mounted on his golden-maned horse, he rode back and forth across the rainbow bridge, ever wary. His teeth were of pure gold, so that his smiles flashed out like sunlight, and his hearing was so keen that he could catch the sound that grass made sprouting fresh and green, and the sound of the wool growing on the back of a sheep. His sight was so sharp that he could see a hundred leagues by night; and he needed less sleep than a bird.

One night, as he kept watch upon Bifrost, he saw a tiny fly flitting in through the window of Freyia's palace. 'What should a fly be doing, entering Sessrumnir by dark?' he wondered. 'It is by day such creatures are wont to be abroad.' And suspicious, he went closer, and watched carefully the doors and windows of Freyia's home.

Now, the fly was Loki who had taken the shape of so small a creature wishing to avoid Heimdall's keen eyes; and once inside Freyia's bedchamber, seeing her to be asleep, he took again his own form, and stealing quietly to the bed where Freyia lay, looked down upon Brisinga-men, glittering in the moonlight with a brighter radiance of its own. But the clasp of the necklace was beneath Freyia's neck on the side that she was lying, and he could not unfasten it without waking her.

Undaunted, Loki changed himself at once into a flea, and hopping on to the bed, he slipped below the coverlet and bit Freyia on the side that was uppermost. She stirred in her sleep and turned over, but did not wake up; and in a moment, Loki, once again himself, was standing beside the bed. He could now easily reach the clasp of the necklace; and very carefully, with gentle fingers, he unfastened it, and Brisinga-men was in his hands. Then quietly he stole from the room, and like a shadow went from Sessrumnir.

But Heimdall, who had never taken his eyes from Freyia's palace, saw him, and instantly he was upon him. Seeing Heimdall, Loki fled, faster than the wind; though fast as he was, Heimdall was yet faster and caught him up. But as Heimdall put out his hands to grasp hold of him, Loki, with a laugh, turned into a blue flame. Heimdall instantly became a cloud, damp and misty, and sought to engulf the flame and quench it, but Loki changed himself into a bear. 'Not so easily, Heimdall, do you conquer me,' he said, and began to lap up the dew of the cloud.

Then Heimdall also took upon himself the form of a bear, even greater than the other, and he and Loki fought together; two strong white bears, biting and scratching and rolling over and over upon the ground, until the night resounded with their snarling.

And when he saw that Heimdall was worsting him, Loki became a seal, wet and slippery, and slid from the grasp of the bear that was Heimdall. Then Heimdall took upon himself the shape of another seal, even mightier than the other, and once again the two gods fought, biting and barking angry taunts. All night they fought together, but before dawn Loki realized that he was beaten and Heimdall was the stronger. So like a flame he slipped from Heimdall's grasp and stood up as himself once more. 'You win, my friend,' he said, holding out Brisinga-men. 'Take the necklace back to Freyia if you will.'

And Heimdall became himself again and took the necklace from Loki. 'You are cunning, Loki,' he said, 'cunning and treacherous. But you are not cunning enough to deceive me, and you are too treacherous ever to be my friend.'

Loki shrugged his shoulders. 'That is as you think, white one. But it was a good fight for all that, even though I lost.' He smiled, 'And who knows, perhaps we shall fight again one day, and that time you may not be so fortunate.'

But Heimdall did not stay to answer him, for he was hurrying back to Sessrumnir with Brisinga-men, before Freyia awoke, so that she should not miss her necklace and be distressed.

VIII

The Six Gifts

THE wife of Thor, the mighty god of thunder, was golden-haired Sif, one of the loveliest of the goddesses. Her long hair fell about her feet, covering her like a cloak, and was her greatest treasure, the admiration of the gods, and the envy of all the other goddesses.

Sif was, indeed, so vain of her golden hair that Loki, in his love of mischief, thought one day how it would be a fine jest to cut it off. Accordingly, at a time when Thor was absent from Asgard, Loki stole one night to Thrudheim, his palace, and going to the room where Sif lay sleeping, he clipped off every golden tress without wakening her.

When Sif awoke in the morning and found her hair cut short, her distress was great, and nothing which the other goddesses might say could comfort her. She wept unceasingly, her head covered with a veil, until Thor came home, striding into Asgard, well pleased by the number of giants he had slain on his journeying, and shouting out a greeting to everyone he saw.

When he reached his palace and Sif ran not forth to welcome him, he was surprised, and went through his halls calling her. But no answer did he get until he found her weeping in a little room. 'What means this?' he cried. 'Why were you not beside the gate to greet me home? This is no way for a wife to welcome her husband.' But when he saw that she wept, his voice grew softer. 'What is amiss, my Sif?' he asked.

'You will not love me any longer when you know,' sobbed Sif. 'You will think me ugly and of no account.' And she took off the veil and showed him her cropped hair.

Furiously Thor shouted, 'Who has done this thing to you?'

'I know not, my dear lord, but I do fear it for the work of Loki. It is a jest such as he ever loves.'

'If it is Loki, he shall pay for it,' said Thor in a voice which thundered throughout Thrudheim, and snatching up his weapons, he hurried out to look for Loki, muttering threats as he went.

Loki saw him coming, frowning terribly and looking from side to side, as though he searched for someone. 'That scowl is the black cloud which comes before a thunderstorm,' thought Loki, and he made to slip quietly away. But Thor saw him go, and with a roar was after him and had caught him fast. 'Did you cut off Sif's hair?' he shouted.

'Why should I cut off Sif's hair?' asked Loki, trying in vain to free himself from Thor's mighty grip.

'I do not want to know why you should cut off Sif's hair, I have asked you did you cut it off, that is all I want to know. For if you did, then here is an end of your mischief, Loki, for I shall break your every bone and fling what is left of you, piece by piece, across the sky, beyond the very sun.'

'What good would that do to Sif?' asked wily Loki. 'Whether I cut off Sif's hair or another did, it is I, and I alone, who have the cunning and skill to find new hair for her. If you carry out your threats, friend Thor, you will always have a crop-haired wife.'

'Can you find new hair for her?' demanded Thor. 'As fine to the touch, as bright a gold, and as long as that which you destroyed?'

'I can do all that,' said Loki confidently.

'Then make haste and do it, and if you fail, there will be one god less in Asgard, I promise you. Go now and do as you boast.'

'Unhand me then,' said Loki, 'and I will show you what I can do.'

Reluctantly, Thor released him, and in a flash, Loki was gone. He went at once below the earth to Svartalfheim and sought out the two sons of Ivaldi, Idunn's brothers, who were famed amongst the dwarfs for their skill in all smith's work. He asked them if they could fashion for him fine golden threads, like hair, which, when placed upon Sif's head, would grow there, as would real hair; and they answered boastfully, 'We could do more for you than that, red Loki.'

And Loki thought to himself how it might soften the wrath of the other gods towards him if he took back to Asgard with him not only new hair for Sif, but a gift for the Allfather and a gift for Frey as well, and he smiled and said, 'Then fashion for me a ship that shall be the best of all ships and worthy of the gods, and a spear that will always find its mark and will disgrace not even such a warrior as Odin.'

'We shall do those things,' said Ivaldi's sons, and they set to work at once, with a great heaping of fuel on their furnace and a great blowing with their bellows and much hammering of metal; while Loki sat and watched them.

And after a time had passed, the sons of Ivaldi brought to Loki a cap of hair of finely-drawn-out threads of gold, as long as the height of a goddess. 'When this cap touches the head of Sif,' they told him, 'it will grow there as though it had ever done so.' And Loki took the hair from them and stowed it carefully away.

And they wrought with wood and metal, and after another space, they brought to him the ship Skidbladnir, with high-beaked prow and flashing shields along her sides. 'This ship,' they told him, 'will hold all the Aesir and all the Vanir, and no matter her course, she will have ever a favourable wind. Yet when there is no desire to sail in her, she may be folded up and carried in a wallet.' And Loki took the ship from them and folded her and put her in his wallet, and he smiled. 'That is a worthy gift for Frey,' he said.

And the sons of Ivaldi fashioned a shaft of strong ashwood cut from Yggdrasill itself, and hammered red-hot iron and set it on the shaft, and brought to Loki the spear Gungnir. 'This spear,' they told him, 'can never miss its mark, and in the hands of Odin it will be a sure defence against all enemies.' And Loki took the spear and marvelled at it. 'You have done well,' he said. 'Your skill is indeed great.' He rose. 'And now farewell to you. I shall bear witness everywhere that the sons of Ivaldi are surely the finest smiths in all Svartalfheim.'

The two dwarfs grinned their broadest grins in pleasure,

that they had impressed a god with their craftsmanship, but before they could reply to Loki's words, a harsh voice spoke from the shadows around the doorway. 'That is a lie, Loki of Asgard, for the greatest smith in all Svartalfheim is my brother Sindri.'

'Who are you that boast so loudly?' asked Loki; while the two sons of Ivaldi protested furiously against the slight that was put upon them in their own forge.

A hideous dwarf came out from the shadows into the firelight. 'I am Brokk,' he said, 'and I can make good my boast, as you call it.'

'That can you not,' cried Ivaldi's sons, stamping their flat feet in rage.

'However skilled your brother,' said Loki, 'I warrant that he could not make three things to equal this hair, this ship, and this spear, which my friends here have made for me.'

Brokk laughed in scorn. 'He could not only match them, he could surpass them easily.'

'That could he not,' laughed Loki.

Brokk smiled maliciously and leant forward. 'Will you wager your head on it?' he asked.

'You are presumptuous,' said Loki, but he laughed again, good-humouredly. 'I will wager my head on it.'

Brokk rubbed his hands together and grinned. 'Your head against my head,' he said. 'My brother's skill against the skill of the sons of Ivaldi. They have made your three gifts for the gods, now shall Sindri make mine, and we will go together into Asgard and all the gods shall judge between us.'

'Tell Sindri to have a care,' said Loki, 'for with all his skill I do not doubt that he cannot fit his brother's head back on his brother's shoulders when once it is cut off. And off it will surely be before many hours are passed.'

62

The two sons of Ivaldi laughed with glee; but Brokk only smiled unpleasantly and went to seek his brother Sindri.

And after a time Loki followed him and listened outside the forge where Sindri worked, and heard how Brokk told his brother of the wager.

'If you will give me your help,' said Sindri, 'I will make for the gods three gifts that shall surpass by far those made by Ivaldi's sons.'

'I will give you all the help I can,' promised Brokk; and Sindri bade him pile the furnace high with wood and blow hard upon it with the bellows; and when the fire was red-hot he threw on it a pigskin. 'Do not cease to blow upon the fire,' he warned, 'or my work will be spoilt.' Then he went out from the smithy, and Brokk toiled at the bellows with all his might. And Loki smiled and turned himself into a gadfly and flew upon Brokk's hand and stung him. But the skin of the ugly dwarf was thick and tough, and he cared nothing for the sting.

Then Sindri returned and bade his brother cease blowing, and took from the fire the boar Gullinbursti, whose bristles were golden wires and shone like a lamp into every corner of the forge. Then he dropped a bar of gold into the furnace, and once more bidding Brokk cease not from blowing, he went from the smithy. Brokk worked the bellows with all his strength and Loki flew and bit him on the neck. But Brokk only shook his head a little, and ceased not in his blowing.

And when Sindri returned, he took from the fire with his tongs the golden arm-ring Draupnir. Then he laid a bar of iron in the furnace, and for the last time bidding Brokk blow without respite, he left the smithy. And Loki flew and settled on Brokk's forehead, and stung him so that the blood ran down into his eyes and he could not see. Then Brokk ceased from his blowing long enough to wipe the blood away with his hand, and for that second only, the bellows fell flat.

When Sindri returned, he took from the fire the mighty battle-hammer Miollnir and dropped it hissing into a cauldron of cold water. But when he came to examine it closely, he found that it was shorter in the handle than he would have wished, because for a moment Brokk had ceased from blowing. 'Nevertheless, it is a mighty weapon,' said Sindri. He gave the three gifts to Brokk, and Brokk hurried with them to Asgard, going by night, when it was dark, that the sunlight might not fall upon him and turn him to stone.

And at an appointed time, all the gods gathered together to judge between the gifts made by Ivaldi's sons and those which Sindri had wrought; and first Loki handed over his three gifts, telling the virtue of each one.

As soon as he had the golden hair in his hands, Thor rose up and called out for Sif, and at the sound of his mighty voice, echoing through Asgard, she came hurrying from Thrudheim to the council hall, and Thor fitted the cap of hair upon her head, and immediately it grew there as though it were her own.

The gods were loud in their praise of Loki's first gift, declaring that now Sif was even more lovely than she had been before, and delightedly she ran back to Thrudheim that she might admire herself in her own silver mirror.

When Frey took hold of Skidbladnir, his ship, and saw how she would open out into a vessel large enough for all the gods and their steeds as well, he did indeed rejoice. And Odin found no fault at all in Gungnir as he held it in his hands and tested it. 'This is the most perfect weapon I have yet seen,' he said. And Loki smiled at the praise his gifts were having.

Then Brokk laid down his three gifts, giving first to Odin the golden arm-ring. 'This is Draupnir,' he said. 'Every ninth night eight rings of the same weight and form will fall from it, so that when the nights of but one year have passed, you may have filled a treasure store.'

'It is indeed a goodly gift,' said Odin, and placed the ring upon his arm.

Then Brokk gave Gullinbursti, the boar, to Frey. 'You may ride upon him, or he will draw your chariot for you,' he said. 'And you will never travel in the darkness when Gullinbursti carries you, for there will ever be the light from his bristles. And besides all this, he will go through water and through the air as easily as over the land.'

Frey took the boar and marvelled at it, saying, 'Indeed, I know not which of my two gifts pleases me more greatly, for they are both wondrous things.'

And then Brokk took the mighty battle-hammer and gave it into Thor's hands. 'This is Miollnir,' he said, 'and however far it is thrown, it will always return to your hand. And with your strength to wield it, it will ever be a sure protection for the gods.'

Overjoyed, Thor tossed the great hammer into the air as though it had been made of wood, and swung it round about his head until it seemed as though several skulls were in danger of being cracked, and Odin, smiling, bade him desist from his dangerous play.

And when the gods came to decide between those gifts made by the sons of Ivaldi and those which Sindri had fashioned, they found themselves unable to decide between the ring Draupnir and Gungnir the spear, or between Gullinbursti the boar and the ship Skidbladnir; but they all agreed that Thor's hammer was a greater prize than Sif's golden hair, for with its aid they might dwell secure in Asgard, well defended from the giants. And so Sindri was declared to be the winner, and Loki to have forfeited his head.

Brokk shouted with glee, but Loki laughed. 'Come and take me, Brokk,' he called. And instantly, like a flame going out, he had disappeared from the hall, and was away.

'Am I to be cheated of his head?' howled Brokk. 'Is there no justice among the gods?' And he appealed to Thor to bring Loki back to him, and willingly Thor went in pursuit, flourishing his hammer and shouting out to Loki. And after a short time he returned, dragging Loki with him, and flung him on the floor before the dwarf. With a dance of malicious triumph, Brokk took a knife from his belt. 'Now is your head mine, Loki,' he said, and took the fire-god by his red hair and brought the knife to his throat.

'Wait,' said Loki. 'My head is yours, but not any part of my neck, so take care that you touch not more than was in our wager.'

Brokk hesitated, and all the gods laughed at Loki's wit and cunning. 'Indeed, he speaks the truth, Brokk, his neck is not yours to harm.'

Brokk thought carefully and then released Loki sullenly. 'It is true,' he said, 'and so am I cheated. Yet since your head is mine, I may at least still your mocking tongue for a time.'

And he took an awl and thread from his belt and sewed Loki's lips together so that he could not speak; while the gods laughed long at the jest, and none more heartily than Thor, bent double in his mirth and slapping his thighs.

But it was not long before Loki had freed his lips and was as mocking and as scornful as ever he had been.

IX

The Theft of Miollnir

THOR'S hammer, Miollnir, soon became one of the gods' great treasures, counted amongst their most valued possessions as a sure defence against their giant enemies. So it was with much distress that one morning Thor awoke to find that it had gone from Thrudheim. He hurried forth at once to look for Loki, thinking that maybe the god of fire had played another of his tricks; but Loki was as surprised as Thor himself to hear that Miollnir was vanished, and said at once, 'I do not doubt that this is the work of the giants. They will have stolen Miollnir away for their own safety. I will go to Iotunheim and spy for you, and maybe I shall find out where they have hidden it.'

So he changed himself into a hawk and flew northwards into Iotunheim, while Thor waited anxiously for his return. On and on flew Loki, but never a sign of Miollnir did he see, and at last he came to the barren lands of Thrym, the ruler of the frost-giants; and thinking to ask him if he had any knowledge of the hammer, he flew to where Thrym sat upon a little hillock close beside his great house, plaiting leashes out of gilded leather for the hounds that lay at his feet, and combing the manes of his horses.

Loki took his own shape, and the tall giant looked at him. 'Greetings, Thrym,' said Loki.

'Greetings, Loki from Asgard. What brings you here? How is it with the gods?'

Loki sighed. 'It goes but ill with the gods,' he said, 'for Thor has lost his hammer, and I do not doubt that it has been stolen. Can you give me tidings of it?'

Thrym threw back his head and laughed until the tears ran down his cheeks.

'Why do you laugh, my friend?' asked Loki.

And when Thrym could speak again he answered him, 'I laugh because Thor's hammer is hidden nine miles deep below the ground where Thor shall never find it. No, nor any of the gods.'

'Is there no other thing that would please you as mightily as does the hammer?' asked Loki. 'Some other thing which the gods could give to you in exchange for what they prize so highly?'

And Thrym thought, and then he said, 'I have heard much talk of the beauty of the lady of the Vanir, how of all the goddesses she is the fairest, and I have long wished for such a wife. Go you back and tell the gods that if they send me Freyia for my wife they may have Miollnir once again.'

'Small hope is there of getting Miollnir back, if Freyia is

to be the price,' thought Loki. But he said aloud, 'I will go to Asgard with your offer, Thrym, and maybe I shall return with an answer.' And in a moment he was once again a hawk and flying southwards towards Asgard.

Thor was waiting in the courtyard before his palace, pacing up and down and ever glancing with impatience at the sky for a sign of Loki; and as soon as he saw the black dot that was the hawk appear, he hurried forward, calling out, 'What news, Loki? Tell me quickly, what news of Miollnir?' long before Loki could have heard even his mighty shouts.

And when Loki came closer, Thor would not allow him time to take his own shape or even to alight upon the ground, before he was at his questioning, 'Tell me quickly, Loki, have you found where Miollnir is?'

'Miollnir is buried, nine miles below the earth, and only Thrym of the frost-giants knows where.'

'But did you not offer him other gifts in exchange? Did you not demand that he should tell you more?' Thor shook his fists in the air with impatience.

'Give me time to finish what I have to tell you,' said Loki. 'Keep silence for a moment.' And when Thor was quiet, he went on, 'Thrym will only give Miollnir back to us if we will send him Freyia as his bride.'

'Then why are we delaying?' demanded Thor. 'Come, let us away to Sessrumnir and tell Freyia of it, that I may have Miollnir back with no more loss of time.'

Loki alighted on the ground and became himself once more. 'Thor, Thor, where are you hurrying to? Do you suppose that Freyia will consent to marry with old Thrym, or that Od, her husband, will be pleased by the suggestion?'

Thor stopped and stared at Loki. 'I had not thought of that,' he said.

Loki laughed. 'For all that, there will be no harm in asking her. Come, Thor, let us go to Sessrumnir, but not at your pace.'

Together they went to Freyia's palace and found her there, though Od was, as so often, away wandering upon the earth. They told her of Thrym's offer, and immediately she flew into a great rage, so that the roof of her hall echoed with her indignation, and she even tore Brisingamen, her necklace, from around her neck, and flung it on the floor. 'How dare one of the giant race offer for my hand as though I were not among the highest of the Vanir? How dare he speak such words to you? And how dare you hear them? And having heard them, how dare you repeat them to me?'

Thor wished himself a long way off, but Loki waited. smiling quietly, until she stopped for breath, and then he said, 'It were best that this matter be put before the council. and debated by all the gods—yes, and the goddesses—so that you may speak to all of us, Freyia, and make your refusal known.'

So Odin called the gods and goddesses to his council hall, and they discussed how they might win back Miollnir and yet keep Freyia for themselves; but no way could they find to do this.

And then at last Heimdall, the watchman, spoke, 'Since Miollnir is Thor's own weapon, and none knows so well how to use it, it seems to me most fitting that Thor himself should go to the home of Thrym and fetch his hammer back.'

With a shout Thor interrupted him. 'It is easy for you to speak, Heimdall, from your post upon Bifrost. But if I go to Thrym's house—though I am strong, as you all know well—I shall be but one against many, and lacking my best weapon. Willingly would I go alone into the house of Thrym had I any hope of success, but without Miollnir I fear that I should perish.'

'I had not intended, Thor, that you should go as a warrior,' said Heimdall. 'But if Freyia would lend you a robe and a cloak, and a veil to cover your head, Thrym would think that it was Freyia herself come for the wedding feast, and would give Miollnir into your hands, and so might you obtain a victory, even against so many.'

Thor jumped up, shouting in his anger so that the very roof beams shook. 'Shall I, Thor of the Aesir, put on woman's garb and go as a bride to Iotunheim? Why, ever after would all the gods and goddesses, yes, and all the giants too, laugh at me for an unmanly coward.'

'The giants could not laugh, Thor,' said Heimdall, 'when you had smashed their skulls for them.'

But Thor would hear no more of the suggestion, and grew angrier with each moment, though all the gods sought to persuade him to assent to Heimdall's plan.

At last Loki said, 'Perhaps, Thor, you would prefer that the giants should leave Iotunheim in their numbers and come against Asgard, knowing that we lack Miollnir to keep them off.'

Thor fell silent for a moment, thinking. 'Come, Thor,' said Loki, 'the safety of all Asgard depends on you. And if you will go to Thrym, decked as a bride, I will go with you as your handmaiden.'

And at last Thor consented, saying, 'Since it is for the safety of all Asgard, I will go. Though if any laughs at me for it afterwards, he will not laugh long.'

So, still grumbling and protesting, Thor suffered his huge tawny beard to be shaved off. 'It will grow all the thicker and the mightier for it after,' said Loki encouragingly. Then he was dressed in a fine robe, reaching to his feet, with a girdle hung with keys, and a long cloak over it; and above all, a thick veil to cover his head. And Freyia, thankful indeed that it was not she who was to go to Iotunheim, herself hung Brisinga-men around his neck, loath though she was to part with it for any time.

Then, when Loki had been dressed in a plain kirtle and a cloak, with a veil to hide his red hair, the two of them climbed into Thor's chariot which was drawn by two strong goats, and away they sped towards Iotunheim, sparks flying from the hoofs of the goats, such speed they made.

When Thrym saw the chariot coming across the barren plain before his home, and spied by the glinting that one of the figures in it wore Freyia's necklace, he leapt up and shouted to his servants to make ready his halls to receive his bride. 'Great wealth have I,' he said, 'many herds of jet-black oxen with gilded horns, great coffers of jewels, and bags of gold, and rich am I accounted among the giants. Yet rich not only among the giants, but among the gods as well, shall I be counted now, since Freyia has come into Iotunheim to be my wife.'

While his servants strewed fresh rushes on the floor, and set up the benches and tables, Thrym went out from his house to welcome Thor and Loki. He took Thor by the hand and led him to the high seat at the end of the hall, thinking, 'How big and strong her hand is, and how tall a maiden is this Freyia, a fit wife indeed for a giant lord.'

The servants hurried in with food and drink, roasted ox-flesh, salmon, huge loaves, beer and mead; and sweetmeats for the women. Thor, by himself, ate one whole ox and eight large salmon, hurrying great mouthfuls of food under his veil; and when he was offered one of the dainty cakes that had been prepared for the bride, he ate not one, but all of them, and drank three vats of beer. Such an appetite was unrivalled, even by the giant-women, and Thrym wondered greatly at it. 'Indeed,' he said, 'never have I seen a bride who ate more than the lady of the Vanir, or who could drink more deeply than fair Freyia. Is such the custom of the goddesses?'

And Loki, who stood behind Thor's chair, leant his head forward and whispered into Thrym's huge ear, 'For eight nights has my mistress fasted, so greatly has she longed to come to your house.' And Thrym was well pleased at the words.

'I am glad that she has longed to see me,' he said, 'for I too have longed to look upon her face,' and he moved aside a little the veil that covered Thor's head. But when he saw Thor's eyes flash angrily at him, he let fall the veil and cried out, startled, 'Her eyes burn like fire! Why should the eyes of Freyia burn thus?'

And again Loki leant forward and whispered in his ear, 'Her eyes are bright and burning with weariness, since for eight nights she has not slept, so greatly has she longed to come to your house.' And Thrym was once more well pleased by his words.

Then Thrym's sister came forward and stood before Thor and asked, 'Fair Freyia, wife of my brother, will you not give to me a gift, as is the bridal custom? A golden ring from your white arm, or a brooch?'

But before Thor could reply to her, Thrym called out, 'A moment, sister. A gift have I for my bride and I will give it to her now.' He turned to his men. 'Bring in the hammer of Thor that was to be the marriage price.'

And his servants carried in Miollnir, and Thrym took it from them and laid it across Thor's knees. 'There, my fair bride, is my marriage gift. You may send it back to Asgard, if you will. Do I not keep well my promises?'

And Thor's right hand closed around Miollnir's handle, and in his heart he rejoiced to feel it in his grip again. Then rising, he swung the hammer above his head, and felled Thrym to the ground. At once there was an uproar in the hall, as the giants sought to escape the fury that had come among them; but Thor went about, striking down to the left and to the right, and no one escaped from him that day. Of all Thrym's household, there was no one left alive.

And that was how Thor won back his hammer, Miollnir, out of Iotunheim.

X

The Children of Loki

THE gentle goddess Sigyn was Loki's wife, and they dwelt together in Asgard, and much sorrow did Loki cause her by his mischief against the gods, for she ever feared their wrath upon him.

One day he tired of Asgard. 'I will go into Iotunheim and find myself a home among the giants, and take to myself a giant-wife, for the gods have grown tedious to me,' he said. And forthwith he left Asgard and travelled into Iotunheim, caring naught for Sigyn's tears. In Iotunheim he met with Angrboda, a grim-faced giant-woman, and smiling at the contrast which she offered to the lovely goddesses of Asgard, he took her to be his wife, and lived with her in a cave in the heart of a dark wood.

75

There in the wood were three monstrous children born to Angrboda and Loki; Fenris-Wolf, Iormungand, and Hel. Fenris-Wolf was a huge grey wolf with gaping jaws and sharp white teeth, who grew with speed to an enormous size. Iormungand was a serpent, hideous and scaly and ever-increasing in length. And Hel was a giant-maiden; one half of her face and half of her body were fair and comely, but all down the other side, from head to toes, her flesh was cold and livid, as might be the flesh of a corpse, and her eyes were dark and deep and pitiless, and no man might look on them without shuddering.

Time passed, and in their home in the wood, Loki's three children grew, so that at last Odin saw them from his throne that was set in his watch-tower, Hlidskialf, and because there was nothing hid from him, he knew that they would live to cause great sorrow to the gods. And he sighed, and tried what he could to prevent that sorrow. He sent to the wood, to Loki's home, and had his three terrible children brought to Asgard, and he pondered long over what he might best do with them.

Then to Hel he said, 'The abode of happiness is no place for you, nor will it ever be. Therefore begone to lowest Niflheim, and be queen there of all those dead who do not die in battle.' And in the depths of cold and misty Niflheim, he had a palace built for her, with high walls and a strong gate, and there she ruled over the spirits of the dead, all save those who entered Valhall.

And to Iormungand he said, 'The home of the gods, where love and kindness dwell, is no place for you. There-fore begone for ever from Asgard into the deep sea.' And he lifted up the serpent and cast him into the sea which flowed all around Midgard. And there Iormungand grew and flourished, until in time he encircled the whole world,

76

and so remained, his tail between his teeth; and thus was he called the Midgard-Serpent.

But to Fenris-Wolf the Allfather said, 'Since much may be achieved by gentleness, and even savage natures may be tamed, remain you with us in Asgard, and in time you may come to be our friend.'

So Fenris-Wolf remained in Asgard and grew mightier and fiercer with each day, until, of all the gods, only Tyr, the brave god of war, dared go near to feed him. And at last the gods saw that Fenris-Wolf never would be tamed; and Odin, knowing that in the end he was to cause them great destruction, ordered that he was to be bound in chains, that he might do no harm until the appointed day which even the father of the gods himself could not prevent.

So the gods made a strong fetter which they called Laeding, stronger than any ten of them could have broken, all dragging on it at once; and they took this chain to Fenris-Wolf and said to him, 'See, here is a chain which we have made to test your strength. Let us bind it about you to see if you can break it, for your strength seems to us to be most wonderful.'

'You may bind me with your chain,' said Fenris-Wolf, 'and I will give you good proof of my strength, I promise you.'

They bound him, and he stretched himself, and Laeding burst asunder, and all the gods pretended great admiration, and praised Fenris-Wolf, though in their hearts they were troubled.

Then the gods made a second chain, stronger even than the first, and called it Dromi. This chain they took to Fenris-Wolf, saying, 'See, here is a fetter even greater than the last. Let us see you break it also.'

And Fenris-Wolf looked at the huge links of the chain

and said, 'Come, bind me, and I will show you my strength yet again.'

They bound him, and he stretched himself once, he stretched himself twice, and the chain broke apart, and he was free again. And all the gods hid the frustration of their hopes and were loud in their praise of the strength of the wolf.

And it became a saying among the Norsemen, 'To get free out of Laeding' or 'To dash out of Dromi', when anything that was most hard was attempted and achieved.

After their second failure to bind Fenris-Wolf, the gods all met together to discuss what they might next do, and Odin sent for Skirnir who was Frey's servant and his friend, and bade him go to Svartalfheim and ask the dwarfs to make for him with sorcery a chain which no strength might break. So Skirnir went to Svartalfheim, below the earth, and asked the most skilled of all the dwarfs if they could make such a thing for the gods.

'We can indeed,' they answered him, 'though it will not be easy, and the task will take us many days.'

So they forged for him a chain which they called Gleipnir, made of six things which have no being: the roots of a rock, the beard of a woman, the sinews of a bear, the spittle of a bird, the breath of a fish, and the sound of a cat's footsteps. And when the chain was made they gave it into Skirnir's hands, and he marvelled at it, that it was so slender and smooth and like a silken thread to the touch, and yet had such strength in it.

Skirnir returned to Asgard with Gleipnir and was eagerly welcomed by the gods, who immediately went to Fenris-Wolf and bade him go with them to an island set in the midst of a lake, saying that they had yet another chain with which to test his strength.

Willingly Fenris-Wolf went with them, for he had grown greatly since the last time they had tried to bind him, and he had no doubt that he would succeed yet once again in breaking any fetter they brought to test him with. But when he saw Gleipnir, how thin and weak it seemed, he grew suspicious lest there were sorcery in it, and he would not suffer himself to be bound.

And when the gods would have persuaded him, he said, 'I do not trust your chain, it is too slight. If you must bind me with it, then let one of you, as a pledge of good faith, put his right hand into my mouth while the bonds are being made fast.'

And the gods looked at one another, and no one spoke; for there was not one of them there who would willingly have lost his right hand. But when a little while had passed, Tyr went forward and laid his hand between the jaws of the wolf.

Then the gods bound Fenris-Wolf with Gleipnir, and when he was well and truly fast, they bade him loose himself, and the great beast stretched and strove against the chain. But the more strongly that he struggled, the firmer grew Gleipnir; until at last he saw that the gods had tricked him, and that he never would be freed. And the gods rejoiced that he was bound at last, and their danger past for a time; but with a growl like thunder, Fenris-Wolf closed his mighty jaws and bit off Tyr's right hand at the wrist.

Then the gods chained Fenris-Wolf to a rock, that he might remain there until the end of all things came.

And in time Loki grew weary of the cave in the dark wood where he dwelt with Angrboda, and he came out of Iotunheim and returned to his home in Asgard; at which Sigyn rejoiced greatly, to have him with her once again, for she loved him well.

XI

Frigg and the Gift of Flax

FRIGG's palace in Asgard was named Fensalir, and there she
lived, attended by the other goddesses and honoured as their
queen. Among those who served her were golden-haired
Fulla, her handmaiden, who kept the casket which held
Frigg's jewels, and fastened on her golden shoes; Hlin, who
went forth into the world of men to care for those whom
Frigg would keep under her protection; Gna, her messenger,
who rode the swift horse Hofvarpnir; Syn, her doorkeeper,
who kept from Fensalir all those whom Frigg did not wish
to have enter her halls; and Eir, who was skilled in the use
of herbs and simples, and taught their properties to the
women of the northlands.

But Frigg did not always remain in Fensalir, for often she went to Midgard and concerned herself with the lives of men and women, to help them and answer their prayers.

A certain poor peasant once lived with his wife in a little house in a valley. Each day he would go to pasture his few sheep on the mountain which overlooked the valley, and always would he take his cross-bow with him, for sometimes he had the luck to shoot one of the wild mountain goats and carry it home for his wife to cook.

One day he saw such a goat, finer and larger than any he had seen before, and leaving his sheep to graze, he climbed towards the boulder where it stood. But as he came close enough to shoot at it, it suddenly leapt away, higher up the mountain, and then stood once again upon a rock too far away for his arrows to reach. He climbed yet higher in pursuit of it, but it was ever the same, and when he was close to the goat, the animal fled, and at last it vanished altogether from his sight. And when the man looked about him, he saw that in his eagerness to kill the goat, he had climbed to the very top of the mountain, where before him rose up a huge gleaming glacier. Then he noticed that there was an opening in this glacier, as it might have been a door. Boldly the peasant went inside, wondering what he would see there, and immediately he found himself in a cave of shining ice, glittering with a thousand gems.

In the centre of the cave stood a fair and gracious lady, robed in shining white, with a bunch of little blue flowers in her hand; Frigg herself, with all around her, her attendants crowned with mountain blossoms, and smiling.

Awed, the peasant fell upon his knees, guessing her to be a goddess. Kindly she welcomed him, and offered him a gift to take away with him. 'Anything that is in this cavern you may have and take back to your home,' she said.

83

His wondering eyes saw all the glittering gems in the ice, all the gold and jewels worn by the goddess and her maidens, and he considered the riches that any one of them might mean to him and his wife, poor as they were; yet ever his eyes went back to the bunch of little blue flowers, bluer than the sky, as blue as her own lovely eyes, which Frigg held in her hand. And at last, though he knew not why, the man said, 'If you would give to me the flowers which you hold in your hand, goddess, that is all the blessing I would ask of you.'

And Frigg smiled, well pleased by his request. 'They are yours,' she said, 'and so long as they are fresh and flourish, so long shall you live and prosper, but when they wither and fade, so will you sink and die. Take them, and with them the bag of seed that I shall give you. Sow it on the land that is your own, and take good care of the plants that grow from it.' And she held out to him the bunch of little blue flowers. He rose and went to her and took it from her hand, then, dazzled by the brightness of her face, he closed his eyes; and when he opened them again, a moment later, he was alone on the mountain top, a bunch of strange blue flowers in his hand, and at his feet a leathern bag of seeds.

Marvelling, he returned home and told his wife of all that had befallen him, and showed her the flowers and the bag of seed. Angrily she upbraided him, saying, 'Why did you not ask for a ring of gold or a jewel that we could have sold? Little enough is the comfort we shall get from a bunch of flowers and a few seeds.'

But the man said nothing to her angry words, he only put his blue flowers in a safe place; and the next day, before he went to the mountain with his sheep, he dug the little plot of land that was his, and planted the seeds which Frigg had given him.

And in time green shoots appeared above the earth, and he tended them and kept the ground free from weeds, until the shoots grew into tall plants with little leaves and tiny flower buds. And then one day the peasant saw that the flowers were opening in the sun, and they were the same blue flowers as those in the bunch which Frigg had given him; and he was pleased, for the little field of blue reminded him of a memory that he would ever treasure, how he had seen and spoken with great Frigg herself.

But his wife said in scorn, 'Your flowers may look very pretty, but what use are they to us? Why, they are not even fodder for the sheep.'

In time the blue petals in the field dropped and the man sighed; but the seed heads ripened and grew heavy, and he smiled to hear the sound which they made when the wind passed through them, rippling them as though they had been a sea. Then the leaves and stems faded and lost their green freshness, and when they were yellowed for two-thirds of their length, one morning, when the peasant and his wife arose at dawn, they saw Frigg, smiling and kindly, standing with her maidens beside their little field.

'You have tended well the seeds I gave you,' she said. 'And now have I come to show you what you must do with the flax, that it may bring you prosperity.'

And she showed them how to pull up the flax plants and steep them in water until their fibres loosened, and after, how to dry them in the sun and beat them with a wooden mallet and comb them to tear apart the strands. And then she showed the peasant's wife how to spin a thread from them, strong and smooth, and weave with it as though it had been wool; and when the task was finished, the woman had a length of linen, the first that she had ever seen. Then with a blessing on them both, Frigg went away.

85

The peasant sold the linen for a good price, and from the seeds that he had saved from his plants, he sowed more flax the following year, so that again he had linen to sell. And in time he and his wife grew rich, with good food to eat and a large house to live in, wide fields of bright-flowered flax and a store-room filled with bales of linen; and their children grew up happily, in plenty, and learnt from their parents the secret Frigg had taught them.

And in all these years the little blue flowers in the bunch which Frigg had given to the peasant in the glacier on the mountain retained their freshness and did not fade, as she had promised. And though in time the woman died, after a long and happy life, the man lived on, to see his grand-children grow up, and after them his great-grandchildren, all in much prosperity.

But one morning, when he looked at the bunch of blue flax-flowers, he saw that they were beginning to droop and fade, and he knew that the time had come for him to die, and he was content. 'But I would see yet once again great Frigg, before I die, that I may thank her for my happy life,' he said to himself. And bidding farewell to his family, he set off up the mountain to seek her.

He was an old man and the climb was hard, but he did not falter, and at last he stood on the very top before the great glacier. Once again he saw the door of ice, and once again he stepped in boldly; and there, as before, were Frigg and her maidens, smiling and kindly.

'You are welcome,' said Frigg, 'we have awaited you. Come and dwell with us for ever in peace and joy.' And the door of ice closed behind the old man, and he was never seen again by his friends. But his flax fields flourished for the good of his family for many generations.

XII

Geirröd and Agnar

ODIN and Frigg concerned themselves much with the welfare of Geirröd and Agnar, the two young sons of King Hraudung; and seeing one day from the watch-tower of Hlidskialf how a sudden autumn storm arose when the two boys were out alone in their little boat, they caused the boat to reach an island off the coast, where the boys might safely come to land. Here Geirröd and Agnar found no one living save an old man and his wife in a little house, who welcomed them kindly, and bade them stay until the wind dropped and they might launch their boat for home once more.

Now, the old man and woman were really Odin and Frigg who had taken mortal shape to help the boys, and, having a fancy to keep near them for a little longer, they caused the winter storms to set in around the coast, so that the boys could not sail out in their little boat for the danger. So during all the winter months, the two boys remained on the island, kindly treated and cared for by the old man and his wife.

Geirröd, the younger son, became Odin's favourite, and Odin taught him much of the use of arms; and on the long winter evenings as they sat before the fire, told him stories of battle-glory, which the boy heard with eagerness. But the shy, dreamy Agnar was more dear to Frigg, and he spent much time at her side, learning greatly from her of wisdom and courage and gentleness towards his fellow men.

When the spring came, and the harsh winds dropped and the sea grew calm, Odin and Frigg bade farewell to the two boys and watched them row towards the coast. Then, well pleased with their two fosterlings, they returned to Asgard.

But not both the boys reached home, for when the little boat touched the coast of their father's land, Geirröd, ever the quicker, jumped out first, and as he turned to his brother, a sudden wicked thought came into his head, and taking the oars from the boat, he pushed her out to sea with Agnar still aboard. A breeze sprang up which blew her far from the shore, and with no oars to row for land, Agnar and the boat were soon lost to sight.

Then Geirröd returned to his father and told how Agnar had been drowned in a sudden squall, and how he himself had only been saved by the kindness of an old man and his wife living on a little island, where he had passed the winter; and his story was believed by King Hraudung, who saw no reason why his son should be lying.

When some years had passed, Odin and Frigg sat together in Hlidskialf and again they thought on Geirröd and Agnar and looked down on them. 'See,' said Odin, 'my Geirröd is now a great king, while your Agnar is no more than a thrall in the halls that should be his.'

And Frigg saw that it was even so, and Agnar, unknown to his brother, had not drowned, but was among the servants in Geirröd's house. Yet she said, 'It is better by far to be a thrall and kindly-hearted, than a great king and miserly.'

'What mean you by that?' asked Odin.

'It is said,' replied Frigg, 'that King Geirröd is so mean that if too many guests come to his house unbidden, he tortures some to scare away the others.'

'That,' said Odin, 'cannot be the truth. When he was a boy, my Geirröd was ever brave and noble, he would not stoop to such shameful deeds.'

'I warrant you that if you went to Geirröd's halls, you would find it even as I say,' declared Frigg.

'That can I not believe,' said Odin. 'Nevertheless, I will go and prove you wrong.'

So Odin went down to Midgard, clad in a dark-blue cloak with a wide-brimmed hat, and set off for King Geirröd's house. But as soon as he was gone from Asgard, Frigg sent her handmaiden, Fulla, with a warning to King Geirröd, that there was an evil sorcerer even then on his way to harm him, and bade him beware of any stranger at whom his dogs did not bark.

Now, Geirröd was not truly inhospitable, in this Frigg had lied, for she wished to see him punished for his crime against his brother, but when he heard that he was in danger from an evil sorcerer, he ordered his servants to bind and bring to him any stranger whom his watch-dogs failed to attack.

And when Odin, in his dark-blue cloak, came up to Geirröd's gate, the dogs, knowing him to be divine, uttered no sound and slunk away, as though they were afraid. So the servants took Odin and led him before their master, saying, 'Here is a stranger, lord, whom the dogs dared not attack.'

Geirröd looked long at Odin, and then asked his name. 'I am called Grimnir,' Odin replied.

'You are an evil sorcerer, come here to work me ill,' said Geirröd.

'That is untrue,' answered Odin. And though Geirröd questioned him long and searchingly, he would say no more than that.

So Geirröd had him bound to a pillar in his great hall, between two fires piled high with wood, that in the pain of the scorching he might speak concerning himself; and for eight nights he left him bound. And on the ninth evening, while King Geirröd feasted, Agnar the thrall, who should have been a king, stole unnoticed to where the stranger suffered, and in pity, gave him a horn full of ale to drink, whispering that he thought it a shame that the king should treat him thus without a cause.

Later that evening, in the midst of the feasting, Odin suddenly raised his voice in a loud chant, throwing back his head against the pillar, the firelight shining on him, as his words rang through the hall, so that all who were present fell silent to hear him. In his song Odin told of the beginning of the world, and of the gods and their homes in Asgard, and of the great ash-tree Yggdrasill. He told the names of the Valkyrs who chose the slain warriors from the battle-field, and he told all the names of Odin, which he took when he went about the earth, and Grimnir was one among them. Then he cried out, 'Hail to you, King Agnar, for you alone shall rule your father's lands, and this boon I give to you as a reward for the drink you offered me. But you, King Geirröd, shall perish, even on your own sword.'

Geirröd, who with all the others had been listening to Odin's chant in wonder, now grew angry at his words, and drew his sword, meaning to hew down the stranger who dared to insult him in his own house. But Odin called out in a mighty voice, 'Hear me, all you who are here tonight, for I am Odin and this is the truth that I have spoken.'

When he heard these words and saw the great light that shone about the stranger's head, and knew that he was indeed Odin, King Geirröd rose up hurriedly to take him from the fire. But he forgot the sword that lay across his knees, and it slipped to the ground, point uppermost; and in his haste Geirröd stumbled and fell upon his own sword, and it pierced him through the heart, even as Odin had foretold.

Then Odin called forth Agnar the thrall, and having named him for the rightful king, he disappeared from sight, being seen no more in those halls. But Agnar ruled long and justly, and was ever kind to strangers.

XIII

Thor's Journey to Utgard

IN company with Loki, and not alone as was more usual for him, Thor once set out for Iotunheim to seek adventure. He took with him Miollnir, his great hammer, and his belt that increased his strength twofold, and he and Loki travelled in his chariot that was drawn by his two goats.

Towards evening they came to the home of a poor farmer who offered them shelter for the night, not knowing who they might be. But he had not in his house enough meat to feed two guests, and ashamed by his poverty, he apologized for the meagre fare which was all that he could give to the strangers.

Thor smiled. 'There is a remedy for that, good farmer,'

he said. And unharnessing his goats from the chariot, he slew them both. 'There lies meat enough for us all,' he said. 'Flay them and bid your good wife busy herself with the cooking, and so shall there be an ample meal for two hungry travellers and for you and your family besides.'

So helped by his son, Thialfi, the farmer flayed the goats, and his wife prepared the goat-flesh for their supper.

'Lay the hides away from the fire, lest they are burnt,' warned Thor. 'And then, as we eat, let us cast on to them the bones of the goats, taking care that none are broken.'

Soon a fine supper was ready, and they all sat down to eat. There was goats' meat in plenty, well roasted, and new-baked bread; and while the bones were picked clean and tossed aside on to the goatskins and the ale cups were re-filled time after time from the vat, the meal passed with much mirth and enjoyment. Thor with his huge tawny beard and his great appetite made the rafters ring with his mighty laughter at red-haired Loki's sly jesting, until the farmer and his wife thought that never had they entertained two more pleasing guests.

The youth Thialfi, his head a little heavy from the ale that he had drunk, holding a thigh-bone of one of the goats, looked at it longingly and wished that he might break it to eat the marrow, and wondered why his father's loud-laughing guest should take so much care to have the bones kept whole. He sighed at the thought of wasting a marrow such as was not come by every day in a house where there was poverty; and then he looked up to see that the hand-some, red-haired other stranger was watching him with a little smile, from his seat beside the fire.

'And why not, lad?' asked Loki quietly. 'What is there against it? The marrow in that bone would be well worth eating.'

Thialfi blushed that his thoughts had been guessed aright, and he looked away. But a few moments later he raised his eyes again and saw that Loki was still watching him, and it seemed to him that Loki's smile was now a little touched with scorn. 'Am I a child, that I should be afraid to disobey a foolish order?' thought Thialfi, and with a quick glance to see that no one else was watching him, he took up his knife and split the bone and ate the marrow out of it. And from the hearth he heard Loki laugh, very quietly.

In the morning, Thor rose up before any of the others were awake, and taking Miollnir, he touched the goat-hides and the bones lightly with the hammer, and immediately the goats stood up, alive and whole. But then Thor saw that one of them was lame in a hind leg. His angry shouts roused the others. 'Someone has broken a bone of my goat,' he said, and his voice thundered through the little house, as he swung his hammer above his head.

'It is Thor. It is Thor himself, and could be no other,' cried the farmer to his wife. And trembling and fearful, he and his family knelt before Thor and implored his pardon for whatever wrong they might have done. And no one of them was more afraid than young Thialfi.

But Loki stood calmly by the ashes of the hearth and smiled to himself at the uproar.

'Take all that we have, great one, but spare us,' pleaded the farmer. 'It is little enough that I have to offer you, for I am a poor man. Yet will I give you the best that I have in recompense. Take my son Thialfi to be your servant. Though young, he is strong, and no man can run as fast as he. He will serve you faithfully. Only spare us all, I beg of you.'

At their terror and their distress, Thor's anger cooled and he laid Miollnir by. 'I will take your son,' he said, 'and

94

may I find him as loyal and as willing as you declare him to be.' And at his words there was joy in the little house once more.

Leaving his goats and his chariot in the care of the farmer and his wife, Thor continued on his way to Iotunheim, striding off with great steps, with Loki at his side, and followed by Thialfi bearing food for the journey.

After a time they crossed the sea into Iotunheim, and went on eastwards through the giants' land, and when night fell they had not yet found anywhere to shelter. But going on a little way in the darkness, they saw before them the shape of a large building with an entrance at one end, very wide and high. Going through this entrance, they found themselves in a hall, and receiving no answer to their calls, they lay down in the darkness and were soon asleep.

But at about midnight they were all three awakened by a great shaking of the earth, and the walls of the house trembled as though they might at any moment fall. Thor jumped up and cried out to his companions, and together they groped around in the darkness of the hall until they came upon a smaller room leading from it, and here they remained, Thor sitting in the doorway with Miollnir held across his knees. The earth had by this time ceased to tremble, but soon there came a noise as of thunder close by, and this continued unbroken until the dawn.

When it was light, Thor went out from the house and saw, lying on the ground near by, an enormous man, and it was this giant's snoring that had sounded as though it were a thunderstorm, and his lying down to rest that had caused the earth to shake. Thor girded on his belt that increased his strength twofold, and gripping Miollnir tightly, he went over to the giant; but as he approached, the huge man awoke and sat up and looked at him.

95

'Who are you, large stranger?' asked Thor.

'I am called Skrymir,' answered the giant, 'and I have no need to ask your name, for I do not doubt from your appearance that you are Thor himself. Am I not right?'

'I am indeed Thor of the Aesir.'

Skrymir yawned and stretched his arms. 'Good morning to you, Thor,' he said. And then he caught sight of his glove, lying a little way off. 'Why, Thor,' he said, 'what have you been doing with my glove?' And he put out his hand to pick it up, and Thor saw that it was no less than the giant's glove which, in the darkness, he and the others had taken for a house; and as for the little room where they had hidden from the earthquake, it had been the thumb of the glove.

'Whither are you bound, my friend?' asked Skrymir.

'We go to the fortress which is called Utgard,' replied Thor.

Skrymir was silent for a moment, then he asked, 'Since you seem to be journeying even in the same direction as I myself, will you not give me your company along the way, Thor, and you too, red-haired Loki?'

'Willingly,' said the two gods.

So first they settled down to eat their morning meal, Thor and his companions from the bag which Thialfi had carried, and the giant from his own huge sack. And after they had eaten, Skrymir proposed that the others should give him their food to carry as well as his own, and they dropped their bag into the sack and he tied up the sack and slung it over his back, and they all set off together.

At evening they reached a grove of great oak-trees, and here Skrymir suggested that they might shelter for the night, and on the others' assenting, he flung down his sack of food, saying, 'I am tired and I would sleep. Let you share out the

96

food, Thor.' And with that he lay down and was soon snoring like a thunderstorm.

Thor took the sack. 'Even if you have no wish to eat, good Skrymir,' he said, 'we others are hungry.' And he made to open the sack and take out the food. But try as he might, he could find no way to loosen the knots of the thongs with which it was tied; and after a time he became mightily angered at the waste of labour, and taking up Miollnir, he went to where Skrymir lay, and swinging the hammer, smote him on the head. Skrymir rolled over and opened his eyes and asked sleepily, 'What is the matter, Thor? It seems to me as though a leaf has fallen on my head and woken me, or was it you who wakened me to tell me you had eaten and were about to go to rest?'

And Thor was ashamed that his great blow had seemed no more to the giant than the fall of an oak-leaf, and he said, 'We have indeed eaten and are about to lie down to sleep. Good night to you, Skrymir.'

Supperless, Thor and Loki lay down under another tree with Thialfi close by, and none too happy in their minds at the thought of their travelling companion, they fell asleep, for they were tired.

At midnight Thor was awakened by the sound of Skrymir's mighty snoring, and could not go to sleep again by reason of the noise. He bore it for as long as he could, and then arose, and taking Miollnir, went again to where the giant lay, and struck him upon the head an even greater blow than the last that he had dealt him, and Miollnir sank deep into the giant's brow.

But Skrymir only awoke and opened his eyes and said, 'Why, are you there, Thor? An acorn must have dropped from the oak-tree and woken me. It is dark, surely it is not yet time to rise?'

And Thor, ashamed that such a mighty stroke had seemed no more to Skrymir than an acorn fallen from a bough, said, 'It is no later than midnight, Skrymir. There is yet time for us to sleep.'

But Thor slept no more that night. Instead he lay and wondered how he might destroy a giant so powerful that even Miollnir seemed a harmless weapon against him. 'For,' he thought, 'it were best that such a one, who may yet prove an enemy to Asgard, however well disposed he now may be, should not be allowed to escape.' And a little before dawn, hearing from his snores that Skrymir was still sound asleep, Thor once more took up Miollnir and went to where the giant lay. And then, with all his strength, he crashed his mighty hammer down upon the giant's upturned temple, and he saw it sink in, even to the haft.

But Skrymir only sat up and passed his hand across his cheek and said, 'The birds must be nesting in this tree, for I thought I felt a twig fall upon my head.'

And Thor was silent, angry and ashamed that the strongest of the gods had proved so puny in power against one who came from Iotunheim.

Skrymir rose and said, 'Our ways must part here, Thor and red-haired Loki, for I must go north towards the hills, and to the east lies the fortress which is called Utgard, where you wish to go. But before we part, let me give to you a word of good advice. I have marked how you three have whispered among yourselves that I am tall and broad and that my strength is great, but in Utgard there are others taller and broader and mightier in strength by far than I. Therefore it were well if, when you come to Utgard and meet its lord, you do not boast of your small prowess.' He paused and laughed. 'Best would it be, little Thor, if you went not on to Utgard, but turned back here. Yet if go there you must, take care that you heed well what I have said. Farewell.' Then, picking up his sack of food, Skrymir slung it over his shoulder and strode away towards the hills. And Thor and Loki and the farmer's son were not sorry to see him go.

'What say you,' asked Thor, 'do we pay heed to his words and go back, or do we go on, as we had intended?'

Loki, ever ready for an adventure, smiled. 'We go on,' he said, 'for I am hungry, and that Skrymir has taken all our food with him.'

So they went on together, and at midday they came to a plain on which stood a huge fortress, so high that they had to stretch their necks and bend their heads backwards to see the top of it. When they came up to this fortress, they found that it had a massive gate of iron bars which was locked fast. But so tall and wide was the gate, that Thor and his companions were easily able to slip in between the bars and so come into the courtyard. The house door being open, they went into the great hall, and it was indeed a vast room, with a bench running along each side, and on the benches many giants sitting at meat. And at the end of the hall, on the high seat, sat the lord of Utgard himself.

Thor and Loki went boldly up to him and greeted him. He looked them up and down and smiled in scorn. 'It seems, from what I have heard of his appearance, that you must be Thor from Asgard,' he said. 'But, oh, what a tiny weakling seems this great Thor of whom I have heard so much. Come now, strangers, are you skilled at any feats? For no one who is not greater than his fellows at some one thing or another sits down to eat in Utgard. What say you, have you any craft or skill?'

And while Thor hesitated, remembering how small his strength had seemed beside the might of Skrymir, Loki spoke. 'I am hungry,' he said, 'and I have waited long for this morning's meal. I will warrant that there is no one here who can eat more quickly than I, when once good food is set before me.'

The lord of Utgard laughed. 'We shall soon try you, red-haired Loki,' he said, 'to see whether you boast without good reason.' And he called down the hall to one of his servants who was named Logi, that he should match himself against his master's guest, to see who could eat more quickly.

A great trough filled high with roasted meat was carried in and set down in the hall; and Loki threw off his cloak and tossed back his flaming hair, and sat at one end of the trough, with the giant Logi at the other; and at a sign from the lord of Utgard, they both began to eat at once. And each of them ate as fast as he was able, so that they both came face to face in the middle of the trough and all the meat was gone. Loki looked up. 'What think you of that for an appetite?' he asked.

But the lord of Utgard pointed to the trough and said, 'It was not ill done, but see, my servant can do better.' And Loki saw that the giant had eaten not only all the meat off the bones, but the bones and his half of the trough as well.

And all the giants in the great hall laughed, and none so loudly as their lord. And Thor frowned in anger and reddened in shame that the gods had been worsted yet again by those from Iotunheim. But Loki only shrugged his shoulders and said, 'Nevertheless, it was a good meal, and I enjoyed it greatly.'

Then the lord of Utgard turned again to Thor and said, 'Come, Thor, what of your other companion? In what does he excel?' He pointed to Thialfi. 'That youth there, is there anything at which he is more accomplished than any other man?'

'I have never yet been beaten by any other man when the

fleetness of my foot was in the question,' said Thialfi. 'Match me against any of your runners, and I will undertake to win a race on any course you choose.'

The lord of Utgard smiled. 'You must indeed run swiftly if that is so,' he said. 'But let us go out to the courtyard and see you prove your words.'

They all went forth from the house to the huge courtyard, which was as wide as any field, and there the lord of Utgard called out a youth named Hugi and bade him run against Thialfi. 'Let them run three times,' he said.

So they ran together once across the courtyard from end to end. And when Hugi reached the farther wall, he was so far ahead of Thialfi that he turned back and ran a little way to meet him.

'That was well run, Thialfi,' said the lord of Utgard. 'I think that never has there come here a man who could run so fast as you. Yet do you not run fast enough to beat my Hugi.'

They ran a second time, and when Hugi reached the farther wall, Thialfi was little more than half-way along the course.

'You run well, Thialfi,' said the lord of Utgard, 'but I do not think that you will win this race.'

And Thialfi and the giant lad ran a third time; and when Hugi had reached the farther wall, Thialfi was not yet half-way across the courtyard. And a great cheer went up from the men of Utgard, for Hugi the winner.

'The sport has made me thirsty,' laughed their lord. 'Come, let us go back into the house and drink. And perhaps there is some small thing at which you, great Thor, are skilled, that you may prove yourself to us of Utgard?'

'I can drink with anyone,' said Thor, 'and drink more deeply.'

'Well said,' laughed the lord of Utgard. And he called to

his serving-boy to bring his drinking horn. 'If this horn is drained in one draught,' he said, 'it is considered that the drinker drinks well, and well enough if it is drained in two. But no one here is so poor a drinker that he cannot drain the horn in three.'

Thor looked at the horn and thought that though long, it was not so very large, and he was thirsty. 'I shall drink it all off easily in one draught,' he said to himself, and raised the horn to his lips. But though he drank deeply and until his breath failed him, when he came to look into the horn, it seemed as though there were yet a great deal within to be drunk.

'If anyone had told me,' said the lord of Utgard, 'that great Thor could not empty my horn in one draught, I should have called him a liar. But now have I seen it with my own eyes. Yet have I no doubt that a second draught will empty it.'

Thor did not answer him, instead he put the horn to his lips again and drank an even mightier draught than the first. But when he had to pause for breath, and he looked into the horn, it seemed as though he had drunk but very little.

'Come, Thor,' said the lord of Utgard, 'you may be reckoned a mighty drinker among the Aesir, but not so here. Yet surely you will not fail to empty my horn at a third attempt?'

Then Thor grew very angry and he determined to succeed, and he drank for a third time with all his might. But when he paused and looked into the horn once more, though the drink was certainly less, still was there much left to be finished. And Thor handed back the horn to the serving-boy and would not drink again, while the lord of Utgard laughed.

'It seems that you are no mighty drinker when judged by our men,' he said. 'But tell me what else you can do.'

'Such draughts would not be considered mean in Asgard,'

said Thor, 'and there, too, is my strength considered great enough. Try me with some feat of strength, and I will not fail in that.'

'There is a little thing which our youths here find much sport in doing,' said the lord of Utgard, 'and it is such a little thing that I should not dare to suggest that the great Thor should try it, had I not seen for myself that he is less mighty than I had ever supposed him to be.'

'What thing is this?' asked Thor, frowning.

'No more than to lift up my cat from the floor,' said the lord of Utgard. And as he spoke there came into the hall a large grey cat.

Thor looked at it. 'That were easily done,' he said. 'It is but a cat, though it is large.' And he went to it and put his hands below it and made to pick it up. But the cat arched itself as he raised it, and the higher he raised it, the higher grew the arch of its back, so that he could in no way lift it up. And at last, after he had put forth all his strength, he had been able to do no more than raise one of the cat's paws from the ground.

The lord of Utgard said scoffingly, 'That matter has gone as I fancied it would, and Thor is proved to be a very little man beside our youths, that he cannot lift up a cat from the floor.'

Thor was greatly angered, and he cried out, 'Now have you made me wrathful with your gibes, and when I am filled with rage, then is my strength increased. Come, let any of you who will, step forth and wrestle with me, and I will show him that my strength is not to be despised.'

But no one answered his challenge, and the giants all sat in their places and laughed at him.

'Indeed,' said the lord of Utgard, 'it appears that there is no one here who would not consider it a disgrace to match himself against one so weak as you have proved yourself to be. But lest you should call us cowards when you tell this tale back home in Asgard, I shall send for Elli, my old nurse, and she will fight with you.'

And immediately there came into the hall a tottering old woman. 'There stands Elli,' laughed the lord of Utgard. 'Let us see the great Thor prove himself her equal.'

But Thor would not. 'Shall I wrestle with a woman, and she an old crone who has seen many years pass by? I, Thor of the Aesir, the strongest of the gods? Why, I should be shamed for evermore that I put forth my strength against a creature so feeble and defenceless.'

'She has thrown many a good man before today,' said the lord of Utgard. 'And if you hold back, all we here shall think that you are afraid of an old woman.'

Stung by his taunts, Thor went forward and took hold of the old nurse, but the harder that he gripped her, the firmer she stood; and when, in her turn, she took a grasp of him, he rocked upon his feet. And try as he might, with all his god's strength, Thor was unable to keep his stand, and he fell down upon one knee.

Then the lord of Utgard called to bid the game to cease, and Thor arose, discomforted at his disgrace, that he had been unable to withstand the petty strength of one old woman. But the lord of Utgard said, 'Evening is upon us, let us to the feasting, and I warrant that you who are from Asgard will not find our cheer unworthy of your notice.' And he led Thor and Loki to seats close by his own; and much of that night they all spent in eating and drinking the good food and ale that were set before them; so that for a while the two gods forgot how they had shown themselves to be of little account in Utgard.

The next day, after the morning meal, Thor and Loki and

young Thialfi set off from Utgard early, and the lord of the fortress went a little distance with them to speed them on their way. When they came to the place where they were to part, the lord of Utgard asked Thor with a smile, 'How think you that your journey to Utgard has fallen out? Are you content with it?'

And Thor frowned and said, 'Truly, I have gained naught but shame and dishonour from my journey to your home.'

But the lord of Utgard shook his head. 'Indeed you have not, great Thor,' he said. 'Now that we have left Utgard many steps behind, I will tell you the truth. If I have the power to keep you out, never again shall you set foot within my fortress. And I will say this also to you, that had I known how great was your strength, I would not have permitted you to enter there even this once. For you must know that it was I whom you met on your way and knew as Skrymir, and I had prepared against your coming to my fortress certain enchantments. The first was when you would have unloosed my sack which held the food. I had fastened it with iron and a spell, so that no one might undo it. And when you struck me those three blows as I slept, the first and least of them would have been enough to kill me, had it reached its mark. But I had put a mountain between myself and you, even that mountain which you see yonder, and your great hammer struck deep into the rock. If you look, you may see how it has cleft it in three places. So also was it sorcery in my fortress. He against whom you, Loki, did contend, my servant Logi, he is wild-fire which can burn the trough and the bones no less easily than it can devour the meat from them. And when you, Thialfi, matched yourself in speed against my Hugi, it could not have been that you should win, for Hugi is thought, and what can run more quickly? And the horn, great Thor, was also an illusion, for when you

105

drank from my horn, though you seemed to have drunk but little, then it was indeed a marvel to us at Utgard, for the other end of the horn was out in the sea, so how could you empty it? When you reach the sea on your journey home, mark well how the tide-line has dropped lower. And when you tried to lift up my cat, Thor, that too was a wonder, for it was no cat, but the Midgard-Serpent, which lies coiled about the earth. When we saw you raise up one of the cat's paws from the ground, then did we tremble indeed at your might. And that you should withstand in the wrestling with old Elli, my nurse, and sink no farther than upon one knee, that was truly marvellous, for she is no other than old age, and who can resist old age?' After a moment the lord of Utgard went on, 'Now must we part, Thor and red-haired Loki, and I am glad that it is so. And I may tell you before we part, that if ever you come again to Utgard, I shall, with all the wiles and sorcery I know, defend myself and my fortress against you. Yet I trust that we may never meet again, great Thor, for you are very mighty.'

When Thor heard how he had been tricked, he grasped Miollnir and flourished it above his head, very wrathful; but in that instant the lord of Utgard vanished, and was nowhere in sight. 'Let us go back to the fortress and cast it down, stone by stone,' thundered Thor. But when they looked, there before them was the plain whereon had stood the fortress, but Utgard was no longer there, and the wide land was empty.

So Thor and his companions returned the way that they had come, and Thor was very angered; though Loki smiled and shrugged his shoulders, caring little. And when they came to the sea, they saw that it was even as the lord of Utgard had told them, and the waters had diminished with Thor's mighty draughts, and thus were formed the ebb-tides.

XIV

The Cauldron of Hymir

AEGIR, the god of the deep ocean, once went to Asgard for a feasting, and so pleased was he at his entertainment, that he invited all the gods to feast with him, in his palace below the waves, at harvest time. 'But,' he said, 'I fear that the cheer which I can offer you may not be as splendid as your own, for though there is no lack of good meat in my house, I have but a small cauldron for the brewing of mead. Yet perchance you may overlook this, if the food is plentiful.'

The gods laughed, and agreed to drink but little when they went to Aegir's feast; all save Thor, who said, 'Good meat is nothing to me, when I cannot drink my fill. You should find yourself a larger cauldron, Aegir.'

'But where am I to find a cauldron large enough to brew

mead for all the gods?' asked Aegir. 'Had I a goat like your Heidrun, who gave mead instead of milk, then would all be well, or had I the cauldron of old Hymir the giant, then would I have one great enough to satisfy the thirst of all the Aesir and the Vanir, even had each one of them Thor's appetite for drink.'

'If the cauldron of Hymir is all you want to make your feast worthwhile attending,' said Thor, 'then shall you have it, and in good time for the brewing.'

'How shall that be, Thor?'

'Why,' laughed Thor, 'I shall fetch it for you myself.' He looked around at the other gods. 'Which of you will come with me?'

'If my one hand is any help to you, then I will go with you,' said Tyr.

'Well spoken, Tyr, I could wish for no better companion. I warrant we shall have an adventure worthy of us both.'

So, taking Miollnir and his belt and gloves which gave him greater strength, Thor set out for Iotunheim with Tyr, and after a long journey they came to Hymir's house, set near the cold sea in a barren, rocky place, where Hymir's cattle grazed on the sparse grass.

Thor knocked loudly on the gate, but Hymir was out hunting, and in the house were only Hymir's wife and his nine hundred-headed mother. The old giant woman looked at them with her twice nine hundred eyes, and cared not for what she saw, and muttering with all her nine hundred mouths, she returned to her place by the fire. But Hymir's wife received them kindly, and gave them mead to drink; and when she heard her husband's footsteps in the courtyard, she hid them behind a row of cooking vessels which hung from a beam that was slung between two pillars. 'Wait there until I have told him of your coming,' she whispered, 'for my husband has often slain an unexpected guest with but one glance of his fiery eyes.'

When Hymir came in, the icicles rattling on his beard and hanging from his bushy eyebrows, his wife told him that they had two guests from Asgard who were even then hidden behind the cooking pots; and at her words the giant cast such a baleful glance in their direction that the beam split open and the brazen vessels fell to the floor with a great clatter, and all of them were broken save the largest cauldron; and Thor and Tyr stepped forth.

Hymir looked at them. 'You are not welcome,' he grumbled, 'since you come from Asgard, but it shall not be said of Hymir that he turned away a guest.' And he ordered his servants to slaughter three oxen for their supper.

His mighty appetite sharpened by the long journey to Hymir's house, Thor ate, by himself, two of the roasted oxen, leaving but one for the others. Hymir frowned greatly when he saw this and said, 'Tomorrow, if you sup in my house, let it be on what I have caught from the sea in the morning.'

'Willingly,' said Thor. 'For I shall welcome a day's fishing with you, Hymir.'

So the next morning, when Hymir dragged his boat down to the shore, Thor went with him and stood waiting. 'You are too small a person to go in my great boat,' taunted Hymir. 'Wait you on the beach until I return with my catch, for I go fishing for whales.'

'And I,' retorted Thor, 'go fishing for something larger than a whale.'

'Then you will need bait, little Thor. Go, find some first,' jeered Hymir.

Thor went straightway to where Hymir's oxen grazed, and choosing out the largest of the herd, he slew it and cut off its head, and returned to the shore. 'You are not a guest whom many would welcome,' said Hymir in anger, when he saw that his finest ox was dead.

But Thor only answered, 'Now that I have bait, will you take me fishing with you?'

Hymir laughed. 'You would no doubt be frozen in a very little while if you rowed out as far from land into the icy sea as I am wont to do.' And he pushed the boat into the water and climbed aboard.

'That shall we see,' said Thor, and jumped in after him.

Hymir rowed mightily, far out into the ocean, and at last he laid the oars aside. 'Here are my fishing grounds,' he said.

'Let us go farther yet,' said Thor. And seizing the oars, he rowed on, until Hymir himself protested. Then Thor ceased, and Hymir baited his hook and soon he caught two huge whales. 'There is our supper, little one,' he said. 'Now let me see you catch a fish.'

Thor baited his hook with the head of the ox and flung it far into the sea, and after a while he felt a great tugging on the line. With mighty pulls Thor dragged his catch to the surface, while Hymir laughed, 'What have you caught, my friend, a flounder? Or perhaps a sprat?'

But he did not laugh any longer, when, above the sea, close beside the boat, appeared the head of Iormungand, the Midgard-Serpent, caught fast on Thor's hook, thrashing his enormous coils about until the sea was churned into froth and the giant's boat rocked as though at any second it would surely be capsized.

'Quickly, Hymir,' gasped Thor. 'Hold you the line for me, while I strike at the monster with Miollnir, and so shall I be avenged for the grey cat which I could not lift in the lord of Utgard's hall.'

But Hymir was afraid for himself and for his boat, and

taking the knife from his belt, he cut the line, and Iormungand, with a mighty roar, sank back into the sea.

'That was a foolish thing to do,' shouted Thor, furiously. 'I have ever thought the giant-people to be stupid, but never did I think that even one of them could do a thing such as you have done this moment.' And he snatched up Miollnir and swung a blow at Hymir, and missed him as the giant leant aside, while the boat rocked and all but overturned; so that Thor was forced to desist, and flung down his hammer. Then angrily and in silence they rowed back to the shore.

When they reached land, Hymir leapt out on to the beach, and slinging the two whales over his shoulder, he started off for his house; upon which Thor picked up boat and tackle and oars and carried those up to the gate.

That evening as they sat at their meal, Hymir said grudgingly, 'Well did you row today, little Thor, and well did you fish, but I will call no man truly strong who cannot break my cup.' And he held out to Thor the glass goblet from which he drank his ale.

'That will be easy,' laughed Tyr, as Thor took the cup and smashed it down upon the table top. But the cup dented the wood and was not even cracked.

And Hymir laughed, 'Try again, little Thor, try again.'

Angered, Thor rose and flung the cup with all his might against one of the pillars of the hall, but the wooden post fell in splinters, while the glass cup was whole.

And Hymir laughed, 'Perhaps you are not strong enough, my friend.'

Thor dashed the cup down upon the floor, but still it remained unbroken, and he became more angry every moment; while Hymir laughed on, throwing back his ugly head and holding his shaking sides.

Then, unseen by anyone else, Hymir's wife came close

to Tyr, and under pretence of pouring his wine, she whispered, 'Hymir's head is harder even than his goblet.'

'Smite Hymir's skull,' said Tyr to Thor; and with a shout, Thor struck the giant on the head with the goblet, so that it was shattered and fell into a thousand pieces on the floor.

Hymir at once ceased his laughing and stared ruefully at the broken cup, and it was Thor's turn to laugh. Then Hymir said, 'Sorry indeed am I to lose my glass goblet, for it was a rare treasure, but you have proved to me your strength, and I would be willing to give a fine gift to one as strong as you. Ask what you will of me, and maybe I shall give it to you, if I see fit.'

Promptly Thor replied, 'I would have the cauldron in which you brew your mead.'

'Sad was I to lose my glass goblet, and yet sadder would I be to lose my cauldron, but even that will I give to you, if you can carry it away.' And Hymir smiled, for he thought that never would Thor be able to take his huge cauldron from the house.

Tyr rose at once, and going to where the cauldron stood, he tried to lift it, but not an inch could he raise it off the floor, though he tried with all the strength of his left hand. 'Try what you can do, Thor,' he said. And Thor seized hold of the rim of the cauldron, and with a mighty heave, he raised it up and placed it on his head, and with a loud whisper to Tyr, 'Come, my friend, it is time for us to be going home,' he walked the length of the giant's hall and out through the door, across the courtyard and through the gate and away over the rocky plain, followed by Tyr.

And Aegir brewed such quantities of mead in Hymir's cauldron that all the gods were loud in praise of the good cheer in his palace; so that, ever after, at harvest time, they went to his sea-halls for the autumn feasting.

XV

Thor's Battle with the Giant Hrungnir

ODIN once rode into Iotunheim on Sleipnir, his eight-legged horse, who was the best of all steeds. From the gateway of his house, Hrungnir, the strongest of the giants, whose head was of stone and whose heart was flint, saw Odin ride by, and marvelling at Sleipnir's speed, wondered who the stranger might be. 'Who are you, in the golden helmet, who ride so fast?' Hrungnir called out.

'I am Odin from Asgard,' answered the Allfather, pausing in his gallop.

'I had heard much of Odin's horse, how great his speed, and now that I have seen Sleipnir for myself, I know that what I heard was true. But swift as he is, your horse is yet not so swift as my Gullfaxi.'

Odin laughed. 'That is not possible,' he said. 'I would wager my head on it, there could be no horse in all Iotunheim so swift as Sleipnir.'

Hrungnir grew angry and shouted for Gullfaxi to be brought, and leaping into the saddle, he raced with Odin across Iotunheim. Over plains and mountains, over rocks and rivers they went, and as Odin had boasted, Sleipnir proved himself the better steed and gained steadily on Gullfaxi. On and on Odin galloped, followed by Hrungnir, across the borders of Iotunheim and right on into Asgard, and there he drew rein.

Hrungnir caught him up. 'You spoke truly,' he said, 'your horse is the very best there is.' And then he looked around him and saw how he had galloped even into Asgard, the home of his enemies, who now came from all directions to greet the Allfather on his return. 'Is this a trap into which you have led me?' he asked Odin. 'Have I ridden over the sea and across the land and through the sky, only to perish within the walls of Asgard?'

'That may no one ever say of the Aesir, that they have tricked one of the giant race into their home and slain him there,' said Odin. 'You are welcome, Hrungnir, to feast with us today in friendship, and I think you will not find our entertainment wanting.'

So Hrungnir stayed that day to feast in Asgard, and there was no one to condemn his company, for Thor, who alone would not have suffered a giant to enter Asgard, was away, as so often, adventuring.

The gods gave Hrungnir of their best, as though he were no enemy; and for his part, he much enjoyed the feasting. But the good mead given by the goat Heidrun was stronger than any he was wont to drink, and before long he had grown drunk and loudly boastful; and of all the goddesses, only Freyia dared pour for him.

'Give me more, more of this delicious mead, fair Freyia,' he shouted, 'for I vow that I will drink all the mead in Asgard, if I am but given time.' And the more she filled his drinking horn for him, the more boastful he became, until at last his words were more than mere discourtesy to his hosts. 'How small a place seems this Asgard,' he said. 'And the hall you call Valhall, why I could lift it up with one hand and carry it to Iotunheim. And so shall I do one day, for none of you puny gods can stand against the might of Hrungnir, once he is armed for battle. Yes, there will come a time, and not so far away—Give me more mead, Freyia—when I shall slay you all, yes, each one of you, and send Asgard crumbling into dust.' He once more drained the horn, and signed for Freyia to refill it. 'Yes,' he went on, 'each one of you shall I destroy, gods and goddesses alike, save only Sif and Freyia; Freyia because she poured my mead, and Sif for her golden hair. Those two shall I take with me into Iotunheim, to be my wives.'

Seeing him to be drunk, the gods had kept silence at Hrungnir's loud affronts, but just at the moment when he was speaking of Sif, how he would take her into Iotunheim to be his wife, Thor entered the hall, and angry as he would have been at the sight of a giant drinking in Asgard, he was yet angrier at hearing Sif's name on Hrungnir's lips. He strode forward, Miollnir held tightly in his hand. 'What means this?' he thundered. 'Who gave Hrungnir leave to enter Asgard? How should a giant drink with us? And why should Freyia pour mead for him, as though he were an honoured guest?'

With insolence Hrungnir turned to Thor and answered

him, 'It was your Odin himself who brought me here, and it is on his bidding that I drink in this hall. It is unfitting that you should question the Allfather's wishes, Thor who shouts too loudly.'

'You will soon regret that bidding which you so lightly took,' cried Thor, and he swung Miollnir above Hrungnir's head.

But Hrungnir cried back at him, 'Have a care, Thor, for it will be small glory to you to slay me here, an unarmed guest.'

And all the other gods called out to Thor to desist. 'For his words are true,' they said, 'and it would be an everlasting shame to us if he were slain weaponless in our midst.'

So, wrathful though he was, Thor heeded their warning, and laid Miollnir by, glowering all the while at Hrungnir.

The giant laughed and said, 'A greater test of your courage would it be, Thor, if you were to meet me at a place which we both should choose, at a time which we both should name, there in single combat to decide which of us is the better.'

Eagerly Thor agreed to this, and they settled on a place where they should meet to fight, and when. 'Foolish was I,' said Hrungnir, 'to ride to Asgard weaponless, for had I my shield and my battle-hone with me now, there need be no delay to our fighting.'

Then Hrungnir went out from the hall, mounted Gullfaxi, and rode away out of Asgard back into Iotunheim. And when the other giants heard that he and Thor were to meet to fight, they thought that they had much to lose, if Hrungnir perished at Thor's hands; for Hrungnir was accounted the strongest of them all, and it would go ill in the future with those weaker than he, if Thor were the victor.

'Since Thor will bring his squire with him,' they said, 'you too must have a squire, and he must be so mighty in stature that the very sight of him will put fear into the heart of the god of thunder.' But since there was, to their minds, none huge enough in all Iotunheim to be their champion's squire, they fashioned a man out of clay, nine miles high and three miles broad, and gave him the heart of a mare set in the midst of the clay.

Then Hrungnir and the man of clay awaited the coming of Thor and Thialfi, the farmer's son, who was his squire. And when Thor's mighty tread was felt, shaking the earth as he came, Hrungnir took up his shield which was made all of heavy stone and in his right hand he brandished his battle-hone, a huge whetstone of flint, and he glared most horribly, a terrible sight to behold. But the mare's heart in the breast of the clay giant fluttered and quaked, for the man of clay was afraid.

'I have an idea to help you, master,' said Thialfi, and he ran on ahead to where Hrungnir waited. 'Great Thor is coming,' he called out. 'My master, great Thor of the Aesir, comes.'

'I am prepared for him,' said Hrungnir. 'Let him come when he will.'

Thialfi laughed. 'Do you call yourself prepared, giant?' he asked. 'You who wait so ill-protected?'

'Indeed, I am not ill-protected,' said Hrungnir with indignation, 'for have I not my shield which no weapon can pierce and my hone which kills when it strikes?'

'You have them truly, Hrungnir,' replied Thialfi, 'but you hold your shield before you, and what use is that to you, since Thor is coming up from the ground against you? When the earth opens beneath your feet, what use will a shield be over your heart? Can you not hear the rumbling of the earth as it splits apart to let Thor forth?'

And the foolish Hrungnir believed Thialfi's words, and

immediately he stood upon his shield. 'Now am I well protected against all assault from the earth,' he said.

At that moment Thor appeared before him, the sparks of anger flying from his tawny beard, and he hurled Miollnir with all his might; and it was too late for Hrungnir to raise up his shield once more. Instead, with both his hands, he flung his hone at Thor. The hone met Miollnir in the air and they came together with a mighty clangour, and the hone broke in half. One half was shattered into a million pieces, so that all the earth around was evermore covered with flints, and the other half went deep into Thor's brow, so that he fell to the ground. But Miollnir struck the giant full upon his stone head and cracked it wide apart, and staggering forward, he fell dead.

Thialfi, flourishing his sword, rushed full at the clay giant, and since he was made of clay, for all the terror in his mare's heart, he could not run away, and Thialfi cut him down with one sweep of his sword. Then he turned to see how Thor had fared, and saw that he was lying with one of the dead Hrungnir's mighty legs across his neck. Thialfi ran to him and tried with all his strength to lift the limb that Thor might be freed, but he could not move it. He called upon all the gods to come and aid him, and in an instant they were there.

Yet they also could not move the mighty giant's leg, and it looked as though Thor would remain in that place, prisoned for evermore.

But suddenly his little son, Magni, came up and begged, 'Let me try what I can do.' And he took hold of the foot of the dead giant, and little though he was, and young, he lifted it easily off his father's neck; while all the gods marvelled at his might and said, 'When he is come to his full strength, he will be greater by far than us.'

Then Thor rose up and embraced his son, and gave to him

for his own, Gullfaxi, Hrungnir's horse. 'For,' he said, 'you have well deserved so fine a prize.' And all the gods rejoiced at the victory.

But the flint from Hrungnir's hone remained for always in Thor's head, and therefore among the Norsemen it was ever considered an ill thing to cast a hone to the ground, lest thereby the flint that was in the head of the thunder-god was disturbed and caused him pain.

XVI

How Loki Outwitted a Giant

A CERTAIN peasant and his wife lived with their young son in a little house nearby the sea. The man was greatly fond of playing chess, and frequent practice with his neighbours on the long winter evenings had brought him much skill at the game. One day a giant came to the cottage and demanded that the peasant should play chess with him. Glad to match himself with an opponent whom he had not tried before, the man consented. 'But for what stakes do we play?' he asked, for he much enjoyed winning a coin or two or a measure of wheat from his neighbours.

'Whichever of us wins may ask the other for what he will of his possessions,' replied the giant promptly.

And the man thought to himself, 'I am poor, there is but little among my possessions that a giant could covet, yet will he own much that would be of use to me, should the victory be mine.' And he agreed to the stakes, and they sat down to play together.

The game was a long one, and went on well into the evening, but at last the giant won. 'Now must you give to me out of all that you own that which I shall choose for myself,' he said.

'Choose and welcome,' said the peasant. 'It was a good game and I would not have missed it.'

'Give me your son,' said the giant. 'He is all I want of your possessions.'

The unfortunate man heard the words with horror; but in vain he and his wife implored the giant to take all else they owned and leave them their only child, for the giant did no more than laugh at their pleading. Yet at last, to their entreaties, he replied, 'I will leave you the boy for one night longer and tomorrow I shall come for him. And more, if tomorrow you have hidden him so well that I cannot find him, I will renounce my claim, and you may keep your son.' And with that he went away.

Weeping, the peasant and his wife wondered where they might hide the child, but they could think of no place where the giant would not look for him. Far on into the night they thought despairingly, and at last the woman said, 'Let us pray to Odin, he may hear us and come to our aid.'

So they prayed to Odin, and after a time there came a great knocking in the darkness on the cottage door. 'It is the giant returned,' said the woman fearfully. But her husband replied, 'It is not yet dawn, he will not come in the night.' And he went bravely to the door and opened it.

Outside stood a stranger in a grey cloak with a wide-brimmed hat, none other than Odin himself. 'I have heard your prayers,' he said. 'Give me your son and I will hide him for you, and maybe the giant will not find him when he comes in the morning.'

Gratefully the peasant and his wife gave the child to Odin; and Odin changed him into a grain of wheat and hid the grain in an ear growing in a wheatfield close by.

In the morning the giant came and looked once around the cottage. 'You have not hidden him here,' he said, and went outside. He looked all about him, and then, having some strange knowledge, he went straight to the wheatfield, while the peasant and his wife watched with troubled eyes. 'Give me a sickle,' demanded the giant, and the man dared not refuse him.

Rapidly the giant cut the wheat, casting aside each armful save one, and then he flung down the sickle and picked out a single ear of wheat from among the rest he held. Then he plucked off, one by one, each grain, until he held the very one which was the boy. The peasant and his wife wrung their hands in despair; but Odin, taking pity on them once again, blew like a puff of wind and tossed the grain of wheat out of the giant's grasp and back to the man and his wife, where it became once more the boy. 'I have done what I can for you,' said Odin. 'Now must you help yourselves.'

The giant strode over to the cottage. 'That was good,' he said, 'but it was not good enough. You will have to outwit me more cleverly than that, if you wish to keep your son. Tomorrow I shall come again to find where you have hidden him.'

All that night the peasant and his wife wondered what they should do, now that Odin could help them no longer; and at last they prayed to Hönir, the bright god, Odin's brother. And just before dawn, when the woman opened the door of the cottage to see if it was yet day, there outside stood Hönir, like a ray of light. 'Give me your son,' he said, 'and I will hide him for you, and maybe the giant will not find him when he comes.'

Gratefully they gave him the boy; and Hönir turned him into a tiny feather and hid him on the breast of a swan which swam on a stream close by.

In the morning the giant came and looked once around the cottage. 'You have not hidden him here,' he said, and went outside. He looked all about him, and then, having some strange knowledge, he went straightway to the stream and snatched up the swan and tore off its head. Then he plucked off its feathers one by one, while the peasant and his wife watched him with dismay. At last the giant held in his hands one fluff of white down which was the boy, and the father and the mother wrung their hands in despair.

But Hönir took pity on them and blew like a puff of wind and blew the feather into the cottage where it became the boy again. 'I have done what I can for you,' said Hönir. 'Now must you help yourselves.'

The giant came to the peasant and said, 'That too was good, but it was not good enough. Tomorrow I shall come again to find where you have hidden him.'

Once more the man and his wife wondered all night what they might do to save their son, now that Odin and Hönir could not help them; and towards dawn they were still sitting over the ashes of their fire, their faces white and drawn.

The man shivered. 'The fire is dying,' he said. And glad to do something other than think, the woman raked the ashes and laid wood on them. Suddenly she said, 'There is yet Loki.'

'Why should Loki help us?' asked her husband.

'We can but pray to him,' she said.

So they prayed to Loki that he would save their son, and waited in the grey darkness for an answer. And suddenly the wood caught fire and blazed up, and in the light of the flames, there stood Loki in the room. 'Give me your son,' he said, 'and I will hide him for you, and maybe the giant will not find him when he comes.'

They gave the child to Loki, who changed him into the tiny egg of a fish, and went from the cottage down to the sea and hid the egg in the roe of a flounder that swam far from the shore.

When the giant came he looked once around the cottage and said, 'He is not here, he is outside.' And he went out and looked about him. Then, having some strange knowledge, he hurried off to fetch his boat and dragged it down to the shore, while the peasant and his wife watched him with terror. But as the giant climbed aboard, Loki came up to him. 'Take me fishing with you,' he demanded.

'Willingly,' said the giant, and they put out to sea together.

In the middle of the ocean the giant baited his hook and fished; but each fish that he caught he flung overboard again, until he caught a certain flounder, and placing it carefully in the bottom of the boat, he rowed once more for the shore.

On the beach he took out his knife and cut the fish open to take out the roe. Splitting the roe, he looked each egg over until he held the one which was the boy between his finger and his thumb. 'This time have I caught you,' he laughed.

'What have you there?' asked Loki.

'No more than a flounder's egg,' said the giant.

'No one would take such care of a flounder's egg,' scoffed Loki. 'Show it to me.'

The giant moved aside his thumb so that Loki could see he spoke the truth, and in that instant Loki snatched the egg and it became once more the child. 'Go, hide yourself in your

124

father's boathouse,' said Loki. And the terrified boy ran across the beach to the little boathouse, and going in, shut the door behind him.

The giant instantly started off in pursuit, flung open the boathouse door and thrust his head inside, shouting to the boy to come out. But as Loki had thought he would, the giant forgot how small the boathouse was, and he struck his head on a beam and fell down senseless. Loki immediately ran up and killed the giant with his own fishing knife, and the boy was saved.

And ever after, the grateful peasant and his wife considered Loki the greatest of all the gods, for he alone had not said to them, 'I have done what I can. Now must you help yourselves,' but had stayed to see his trick through to the end.

XVII

Odin and Rind

THERE came a time when Odin, with his power of seeing those things which were yet to come, grew greatly troubled; for, looking forward into the future, he saw how one of his sons was doomed to die by the hand of another, bringing much grief to Asgard and all the gods. But because nothing was hidden from him, the Allfather saw also how another son of his would be born who would avenge his brother's death. The name of this son would be Vali, and his mother would be Rind, a princess in an eastern kingdom of Midgard. So, heavy at heart for all the unhappiness that was yet to come, Odin went down to earth and set himself to woo and win the princess who was to be Vali's mother.

Rind was very beautiful, and exceedingly proud; and one after another she had rejected all her suitors, saying that they were neither great nor noble enough to wed her, until her father despaired of ever finding her a husband.

But he soon had other troubles beside his daughter's wilfulness, for an enemy gathered together a great army and marched against his kingdom; and being old, the king knew himself to be but a poor warrior, though he had been accounted of much prowess in his youth. He sat upon his throne and thought over all his captains, wondering to which of them he might give the command of his army, and in his mind he doubted the abilities of each one.

While he sat, thinking unhappily, a stranger of middle age, wearing a grey cloak and a wide-brimmed hat and having but one eye, came into the hall and spoke to him. 'Give me the command of your army,' he said, 'and I will win the battle for you.'

The stranger's words were boastful, but there was something in his bearing that made the king believe them, and it was without misgivings that the old king sent forth his men against the enemy with the stranger at their head.

And Odin, for it was he, made good his words, and the enemy was utterly destroyed. Overjoyed, the king welcomed back the stranger, saying, 'Any reward that you claim is yours. Ask what you will of me.'

'All I ask,' said Odin, 'is leave to woo your daughter.'

'It is yours,' said the king, 'and may you be successful.' And he thought how from gratitude at least, Rind might look kindly on the stranger who had saved her father's kingdom.

But when Odin went to Rind and declared himself her suitor, never a smile or a word of thanks did the lovely maiden offer him. She did no more than look him up and

down and frown a little. 'The man I marry must be young,' she said, 'and you are of middle years.' And she turned away and spoke no more to him.

Odin went from the king's house, and the old king was sad to see him go and angry that Rind should have slighted him. But before many days had passed, Odin returned in the guise of a worker in metals and precious stones, young and handsome.

The king welcomed him for his craftsmanship, and he made for the king rings and brooches and silver cups, while for Rind he made necklaces and other jewellery which gave much pleasure to her. Well pleased by his skill, the king wished to reward the young smith for his work. 'What would you ask of me?' he said. 'For I would give you a gift.'

Odin smiled. 'There is nothing I would ask of you, save permission to woo your daughter.'

'It is yours,' said the king, 'and may you be successful.'

But pleased as she had been by his skill and by the jewels he had wrought for her, Rind had no smile or word of greeting when Odin came to her; she only looked him up and down and said with scorn, 'You are young and handsome, yet you are no more than a smith. The man I marry must have proved himself a brave and noble warrior.' And she sent him away from her without another word.

The king was sad to see him go, for he feared that there would be no pleasing his daughter, and that in the end she would die unwed through her great pride. But before many days had passed, Odin came again to the king's house, this time as a warrior, young and bold, rich, noble, and handsome. He greeted the king with respect and begged that he might be allowed to woo his daughter.

The king's heart lightened at the sight of Rind's latest suitor. 'Surely,' he thought, 'no maiden could refuse so fine a man.' And eagerly he welcomed Odin. 'Go to my daughter, stranger,' he said, 'and may your wooing be successful.'

'I have long admired your beauty, fair Rind,' said Odin, 'and I have come from far away to ask you to be my wife.'

Rind looked him up and down and she hesitated, so that her waiting-women glanced at one another and smiled, thinking, 'This time her hand is won.'

But at last Rind spoke. 'Stranger,' she said, 'you are handsome, and your garments show you to be rich; you seem to me to be a mighty warrior, and your bearing proves you noble. Such a one as you would be a worthy husband for any maiden save myself. But my beauty is not for any mortal man.' And she turned her head away from him.

Then Odin grew angry. 'This time, fair Rind, you have scorned me once too often.' In a voice so quiet that only she could hear him, he spoke magic runes over her, and with a cry she fell senseless into the arms of her attendants. And flinging his cloak around him, Odin strode from the room.

The king's distress was great, for he thought that Rind was dying, but after a time she recovered her senses. Yet it seemed as though she had lost her reason, for she no longer smiled or laughed, she would speak to no one, and, day after day, would do no more than sit staring before her at nothing at all.

In vain the king called to his house all the women of his country who were known for their skill in medicine, for no one of them could tell him what ailed her. 'She has angered the gods,' was all they could say, 'and the gods alone can help her.'

And then one day Odin came again to the king, and this time he was in the shape of an old woman, poor and ragged. 'I can cure your daughter,' he said to the king. 'Let me but speak with her alone.'

So the king led the old woman to the hall where Rind sat

and he dismissed all her attendants. 'There is my daughter, alone and in your power, old woman,' he said. 'If you can cure her, there is nothing you may not ask of me.' And he went from the room.

'Rind,' said Odin quietly, 'would you be satisfied with a husband from among the gods?'

Rind looked at him suspiciously. 'What can you know of my wishes, old woman?'

'Fair Rind, you scorned me as the captain of your father's army, you despised me as a smith, you rejected me as a warrior, what will you say to me as a god?' And Odin took his own true shape, bright and shining and glorious to look upon, and in terror Rind fell upon her knees and besought him to forgive her.

Odin smiled. 'You have given me no answer yet. Will you be the wife of a god?'

'Willingly,' she whispered.

'It will be only a short time that we can be together,' Odin warned her gravely, 'since before many days are passed I must return to Asgard, and you may never see me again. Yet you will be held in respect above all women and a son shall be born to us who will do great deeds. Think you the honour worth the price?'

Rind rose and looked into Odin's eyes. 'I have ever known,' she said, 'that I was destined to great glory.'

So Rind was married to Odin, yet he dwelt with her in her father's house for no more than a little space, as he had foretold. And one day he said to her, 'I must leave you now, fair Rind. When our son is born, name him Vali, and send him forth to find his father, for there will be a task for him to perform. Farewell.' And before she could reply to him or seek to keep him with her, he was gone, as though he had never been there.

XVIII

The Death of Balder

BALDER, the beautiful young sun-god, was the happiest of all the gods, for ever gay and joyous, unlike his melancholy blind twin brother, Höd, the god of darkness; and he was the best loved of all who dwelt in Asgard. But there came a time when Balder grew pale and sad, as though something troubled him, so that Nanna, his beloved wife, seeing him so downcast, asked him what was amiss.

'It is,' he replied, 'that I am distressed by a dream which I have had, night after night, of how my life is to be taken, by whom I know not, and no one of all the gods may save me from the house of the dead, where grim Hel rules.'

Grieved at his words, Nanna went to Odin and Frigg and told them of their son's dreams, and they heard her gravely. 'We must call the gods and goddesses to a council, to decide what should be done for Balder's safety,' said Odin, 'for such dreams can bode no good to him.'

But before he went himself to the council, Odin climbed to his throne on his watch-tower, Hlidskialf, and looked out over Midgard and Iotunheim and down into misty Niflheim even to the house of Hel, Loki's terrible daughter. And he saw that her halls were swept and garnished, her high chairs hung with coverings, and the cups set out on her tables as for a feast; and he knew that her house awaited the coming of a greatly honoured guest. 'This,' thought Odin, 'is the beginning of that end which is to be.' And with a heavy heart he joined the other gods at their council.

After they had debated long, it was decided that Balder's safety could only be assured if oaths were taken of all things that they would not harm him. And forthwith Frigg herself sent out her messengers to ask of each thing there was that it might promise no hurt to her son. Fire and water, the trees, the plants and the flowers, stones and rocks, the earth and the metals beneath the earth, sickness and plagues, birds and all animals, each one swore to Balder's safety. And when that had been done, all the gods save only Odin smiled and said, 'Now can Balder be harmed by naught.' But Odin still feared, though he spoke not of his fears.

And it became a great sport with the gods to cast at Balder all manner of weapons, and sticks and stones, to see him quite unharmed by them; and they would laugh greatly and rejoice, for they held Balder very dear.

But Loki alone of all the gods did not rejoice, since fire must ever be jealous of the sun, which is more bright and better loved, and Loki longed that Balder's beauty and light might be put out for ever, and he sought a way to harm him. In the likeness of an old woman he went to Fensalir, where Frigg sat spinning the clouds, and he greeted her.

'Have you passed by the courtyard where the gods are gathered?' asked Frigg. 'What do they there?'

'They cast stones and weapons at Balder, your son,' said Loki, 'and he stands unharmed.'

Frigg smiled happily. 'Each thing there is has sworn to me that Balder will be safe from it,' she said.

'Are you certain that each thing has sworn?' asked Loki. 'Is there no thing at all that you have forgotten?'

'No thing at all has been forgotten,' said Frigg. 'One thing alone has not sworn, though I did not forget it. It is the mistletoe that grows on the oak-tree that stands at the gate of Valhall. It is such a little thing that it could harm no one, so I did not think to take an oath from it. It will do no hurt to Balder.'

'Indeed, why should it harm Balder?' said Loki. 'Do not all things love him?' And bidding Frigg farewell, he went from Fensalir. Immediately, taking his own shape, he hastened to the oak-tree that grew at the gate of Valhall and cut the little branch of mistletoe and fashioned it into a dart. Then he went to the courtyard where the gods were gathered, with Balder in their midst, following their favourite game of throwing weapons at him, while he stood there laughing and unharmed.

Only blind Höd waited alone, apart from the others, because he could not see to aim a weapon or a stone. To him Loki went and asked, 'Why do you not play with the others, Höd? They have great sport.'

Bitterly Höd answered him, 'It is always dark where I am, so how can I see the sun?'

'Here is a shaft,' said Loki, putting the mistletoe dart into his hand. 'Turn around, and I will guide your aim.'

Höd smiled. 'You are kind,' he said. 'I am always lonely because I cannot join the others in their games, but now

135

you have lent me your own eyes.' And he took aim as Loki guided him and cast the little dart. And it pierced Balder through the heart so that he fell down dead; and with a smile, Loki stole away.

The gods' gay laughter died, and a deep silence fell; and in the silence Höd asked, 'What is amiss that no one speaks?' And someone answered him, 'Balder is dead and you have slain him.' And all the gods fell to weeping. With a cry, Höd ran from the place, stumbling and afraid; and going for ever from Asgard, sought the shelter of a dark forest, where he groped his way among the trees, always listening lest the others might come after him to destroy him for his unwitting crime.

Frigg came from Fensalir to see how her son was slain, and bitter were the tears she wept for him. After a time she spoke and said, 'If there is any one among the gods who would have my love for evermore, let him go down to the house of Hel and offer her a ransom, that Balder, my dear son, may be returned to us. For without Balder our days are dark indeed.'

And bold Hermod, the messenger, stepped forth and said, 'I will go down into the house of Hel and speak with her.'

Odin gave to him his own horse, Sleipnir, to carry him along his dreadful road, and sped by the hope of all the gods, Hermod rode away to misty Niflheim.

The gods carried Balder's body to the sea and laid it on his own ship, Hringhorni, and placed his armour and his sword by him, and heaped all around him resin-scented pine logs. But when the bale-fire was built, they found that they could not launch the ship, for she was too heavy. So the gods sent into Iotunheim for the help of the giant-woman Hyrrokkin, and she came riding on a huge wolf bridled with snakes. She went to the ship, and with one mighty thrust, she sent her down the beach into the sea.

Then, before all the assembled gods, and before many of the giants who had come from Iotunheim to bid a last farewell to Balder, whom all, gods and men and giants alike, had loved, save only Loki, Odin took up a flaming torch and went to fire the pile. But as she watched, Nanna's sad heart broke when she thought how she would never see her husband more, and she died. So the gods laid her gently on the pyre beside Balder, and Odin placed by them his arm-ring Draupnir, that Sindri the dwarf had made for him, and he thrust his torch among the wood, and the flames blazed up, as the ship moved slowly out to sea. And from the shore the gods watched her, weeping, until she was quite consumed by fire.

For nine nights Hermod rode through the darkness until he reached the bridge that spanned Gioll, the river that flowed through Niflheim, and Modgud, the maiden who kept the bridge, called out to him, 'Stay, and tell me your name and your errand, you who ride by so fast. Yesterday did five companies of dead men pass this way, yet they made no sound as they crossed my bridge. But your horse's hoofs have a noise like thunder, and your face has not the colour of death. Say who you are and what is your will.'

'I am Hermod from Asgard, and I seek out Hel, that I may ransom Balder. Has Balder passed this way?'

'He has passed this way, he and his wife, and they are even now in the house of Hel,' said Modgud.

So Hermod rode on until he saw the walls of Hel's house before him, and her gate, with Garm, her huge hound, guarding it. But there was no one there to open to him, so he rode Sleipnir at the gate, and Sleipnir leapt right over, into the home of Hel.

Hermod rode across the dark and silent courtyard, and dismounting, went into Hel's great hall where she sat upon her throne. Close beside her, in that dread place, sitting in silence at the high table, hand in hand, were Balder and Nanna, pale and dim, with the light all gone from Balder's face. And they both arose and embraced Hermod with joy, and greeted him.

Then Hermod knelt before Hel and asked her to name a ransom for Balder. 'For,' he said, 'the gods will never have done weeping for him, and all things sorrow at his death. Release him, mighty Hel, that he may return to us.'

And Hel looked at him with her deep, dark, sad eyes, that no one might see without shuddering, and in her harsh voice she replied, 'If Balder is so truly beloved as you tell me, then may he return to life. But if there is found one thing that will not weep for him, then he remains here with me.'

With hope Hermod bade farewell to Balder and Nanna, saying, 'May it be soon when we meet again, and in joy.' And Balder took from his arm the ring Draupnir and gave it to Hermod. 'Take this to my father from me,' he said, 'as a remembrance.'

And to Frigg Nanna sent the robe she was wearing, and a golden ring from her finger to Fulla, Frigg's handmaiden. And at the gateway of Hel's house Balder and Nanna stood and watched Hermod ride off on Sleipnir, back to life and Asgard.

Hermod told the gods of all that he had seen and heard in the house of Hel, and immediately messengers were sent out to bid all things weep, that Balder might come back to them. And all the things there were wept; the gods, the giants, the dwarfs and the elves, all men and women, the flowers and the trees, even the rocks and the stones, all things in every place; so that everywhere the sound of lamentation filled the air. But as the messengers were hastening back to Asgard, believing their task completed, they passed by a cave where a giant-woman sat whom they had not seen before. 'Who are you?' they asked her.

'I am called Thökk,' she replied.

And they bade her weep Balder out of the house of Hel, and she answered them, 'Why should I weep for Balder? I loved him not. Let Hel keep what she holds.' And she laughed and was gone.

Sadly the messengers returned to Asgard and told how of all things one giant-woman only had not wept for the god of the sun. And they told how to their ears, her laugh had seemed like the mocking laugh of Loki, who was so skilled at changing his shape. And there was great grief in Asgard, that Balder had to remain in the house of Hel, with his death yet unavenged.

But on a certain day there came a youth to Asgard and demanded admittance to Odin's hall, and the doorkeeper would not let him pass. 'I am awaited,' said the boy quietly, and he thrust the doorkeeper aside and went through the hall to where Odin sat. Standing before Odin, young and proud, with his bow in his hand and a quiver of arrows slung at his shoulder, he said, 'I am Vali, and my mother Rind has sent me to you, for you have a task for me to perform.'

Odin smiled sadly. 'You know your task,' he said. 'Go now and do it.'

And Vali turned and left the hall, while the gods marvelled at him. He went to the wood where Höd wandered fearfully, and going swiftly among the trees like a ray of light, he fitted an arrow to his bow. And Höd heard his footsteps and knew that the end was near for him, and he was not grieved, for his life had grown weary. Vali's arrow sped true and found its mark, and Höd went down to the house of Hel, where his brother Balder greeted him in love and kindness. And thus did all things come to pass as Odin had known they would.

XIX

How Loki was Cast Out by the Gods

Though the gods still mourned for the loss of Balder, when the time came round again when they were wont to go to feast with Aegir in his palace under the sea and drink the mead which he brewed in the cauldron of Hymir, they went as was their custom, all save Thor, who was journeying in the north, in Iotunheim, for they wished to do courtesy to Aegir. In Aegir's halls, lighted by the glittering gold that shone upon the walls in place of torches, the gods set themselves to forget their sorrow and enjoy the entertainment offered them, eating the good meat put before them and drinking the abundant mead; and there was great peace among them.

But while they feasted and told tales of their feats of arms, Loki, who had been apart from them since Balder's death, came to the door of the hall and made to enter. Eldir, Aegir's serving-man, would not let him pass, saying, 'These are no friends of yours gathered here today, Loki. Go you back whence you came.'

'I have come to feast with the other gods as is my right,' said Loki, 'whether they wish my company or not.' And he went on into the hall.

And when the gods saw that he had come among them, they all fell silent, and there was not one of them who was glad to see him there; not even Sigyn, his wife, for she was afraid for him.

Loki looked around the hall and laughed. 'I am thirsty,' he said, 'and I have journeyed far to drink of Aegir's mead. Why are you all so proud and silent? Why do you not bid me sit and drink with you?'

And Bragi, the god of poetry, answered him, 'Your place is no longer with us, Loki, for you have forfeited the right to our friendship by your evil deeds.' And the other gods murmured their agreement with his words.

Loki turned to Odin. 'Do you remember, Odin, how once we took an oath to be as brothers in all things? There was a time when you would not have drunk the ale that was poured for you, unless it were poured for me also.'

And because he spoke the truth, Odin could not gainsay him, and he turned to the others and bade them make room at the table for Loki. 'The father of Fenris-Wolf would sit and drink with us,' he said. 'Bring a cup and mead for him.'

Odin's son, Vidar, the silent god, rose and gave his place to Loki and poured out mead for him. Loki took the crystal cup and raised it. 'Greetings to you all, brave gods and fair goddesses. I wish you well, all save Bragi yonder, who would have sent me away.' And he laughed and drank.

Quietly Bragi answered him, 'I would give much, Loki, if I might prevent strife and quarrelling here this day.'

Loki laughed again. 'I do not doubt it, Bragi, for of all the Aesir and the Vanir who are gathered here, you are the least valorous in battle, and ever ready to avoid a quarrel.'

Angrily Bragi answered him, 'If we were not guests in Aegir's hall, your head would pay for your lies.'

But Idunn, his wife, laid her hand upon his arm and pleaded, 'He is Odin's sworn brother, Bragi, speak to him gently. Harsh words avail nothing.'

'Hear how Idunn from Svartalfheim seeks to keep the peace that her Bragi may not be called upon to fight,' mocked Loki.

'Come, Loki,' said Odin, 'this is no place for wrangling. Drink and be at peace with us while you may.'

'Are you afraid, Odin, that I shall tell aloud of how you have not always been wise or just when you wandered amongst men? Or of how you worked as a thrall for a giant, when you reaped Baugi's fields for him?'

'What do old tales matter to us here today?' said Frigg gently. 'Forget what you and Odin may have done in the past. And if it is folly that is forgotten, it is so much the better for that.'

Quickly, Loki turned to Frigg. 'I do not doubt, Frigg, that you would prefer that I should forget also how you cheated Odin once when you sent a lying message to Geirröd the king, that you might not lose a wager you had made.'

'If my dear Balder were still with me, no one would dare to speak to me thus,' said Frigg bitterly.

'You have me to thank that he is not, good Frigg.' Loki smiled and raised his cup to his lips.

'You are out of your mind, Loki,' exclaimed Freyia, 'that you boast of your shameful deeds here, before us all.'

142

Loki put down his cup and turned to Freyia. 'I did but speak a few true words, Freyia, where is the shame in that? You, I see, are wearing Brisinga-men, here, before us all. Do you remember how, in Svartalfheim, you paid with kisses for your necklace? You, the lady of the Vanir?'

Freyia blushed and looked away confused, having no answer that she could give him; but Niord, her father, spoke for her, 'However many slanders you may speak against fair Freyia, Loki, at least you can never say of her so strange a thing as that she was the mother of an eight-legged foal.' And all the gods and goddesses laughed at his words, and even Loki himself smiled, for he never could resist a jest. 'And of Frey, my other child,' went on Niord, 'surely not even Loki's malice can find evil to speak.'

Tyr, the god of war, who had until then been silent, now spoke up in praise of Frey. 'Indeed, Niord, he is the best of all gods, brave and gentle and kindly. A good friend is Frey.'

At once Loki sneered at Tyr. 'A good friend! Maybe the words fit Frey, but a good friend is a name which even your friends could never give you, you whose chiefest delight it is to stir up strife and bloodshed, you who are ever near where there is battle and slaughter. Surely there could be found some more peaceable occupation for a god who lacks a right hand to hold a sword?'

'I may lack a hand,' answered Tyr angrily, 'but you lack your son, Fenris-Wolf, who stays in chains until the end of all things comes. And there is no one here save you, Loki, who would think to mock me that I must wield my weapons with my left hand because I once did a service for Asgard.'

Just as Tyr had spoken for him, so now Frey spoke for Tyr. 'Tyr is the bravest of us all, Loki, and you do rashly to speak so to him. Beware lest you yourself join your monster-child in chains.'

Loki smiled. 'How foolish you are, Frey, to provoke my words, it would have been wiser to have remained silent. Do you wish that I should remind you of how you once sat on the Allfather's throne and saw a certain giant-maid and grew sick with love for her? Unfortunate Frey, she would not receive your gifts, and at last you paid for her love with a sword which you gave to Skirnir. When the end of all things comes, and you have no sword to defend yourself, you will regret that you gave your sword to win yourself a wife.'

At this Skirnir called out in defence of his master, 'Were I great Frey, I would not wait to hear further evil from the lips of Loki, that speaker of ill. I would break and crush him until he was no more.'

Loki drained another cup of mead. 'It seemed to me then as though I heard a voice which spoke my name. What little creeping creature spoke of Loki?'

'I am Skirnir and proud to serve great Frey. And proud too am I that because of my master the gods permit me to drink with them today.'

'Ah yes,' smiled Loki, 'Skirnir is your name. I have noticed you, following after Frey like an obedient dog, and running errands for him.' And from shame and anger Skirnir did not answer him.

'Loki, you have drunk too deeply of Aegir's good mead, and you do not know what words you speak,' said Heimdall, who sought to prevent more quarrelling.

'Why, Heimdall, are you, too, here today? Have you then grown weary of watching from your post upon Bifrost? A good watchman never leaves his post to go feasting with his friends.' Loki laughed. 'Do you remember, Heimdall, how we once fought for Freyia's necklace?'

Skadi, who had been silent ever since Loki had entered the hall, now could no longer refrain from speaking; frowning and angry, her voice bitter with hatred for Loki whom she blamed for the death of her father, Thiazi, she called across the hall from where she sat. 'Beware, Loki, for one day an evil fate will come to you, and when that day arrives, I shall smile and be glad.'

'Fair Skadi,' he answered her, 'no one will prevent you if you wish to smile and be glad at my ill fortune, yet all your smiles and gladness will not bring back to you Thiazi, the old eagle. There is naught left of him now save two stars set in the sky.'

'And I shall ever hate you for that, Loki,' she said in a low voice. But he only laughed, and looked into his cup which was empty again.

Then Sif, Thor's wife, thinking how his mocking tongue had spared her until then, and seeking to win his favour with soft words, rose and came to Loki with a jar of mead, and smiled at him, and filled his crystal cup and said, 'Drink, good Loki, and be friends with us. Of almost all who are gathered here today, have you spoken ill words, yet in me you have found no fault; for I am blameless, am I not, good Loki?' And she smiled again at him.

Loki took the mead she gave him and drank, then he set down the cup and looked at her and laughed a little. 'If I have spared you, Sif, it was because I had forgotten you, or did not notice you among so many fairer goddesses. For all your sly smiles, Sif, and the mead you pour for me, you are no better than the others here. Shall I tell them some tales of you, that you may hear them laugh?'

But before Sif could reply, there was a heavy tread without and a shadow at the doorway, and Thor entered the hall, followed by Thialfi, and with a cry Sif ran to him for his protection. 'What means this,' demanded Thor, 'why drinks that traitor with us? He would not have dared to speak Sif's name had I been here when he began his slanders. Have a care, Loki, lest Miollnir break your skull for you.'

'Your threats were ever loud, Thor, but I do not think that you will find much time to talk when the end of all things comes, as come it shall. Then you will have to act, not threaten.'

Thor thrust Sif aside and strode forward. 'Be silent, Loki, or I will pick you up in one hand and fling you northwards, even into Iotunheim, where you belong.'

Loki laughed. 'If I were Thor, I would not speak of Iotunheim. Do you remember, Thor, that time we travelled northwards and slept in a glove? You sat all night, afraid, with Miollnir nursed in your arms, like a woman with her child, waiting to be attacked by unknown enemies. And it was no more than a giant's snoring that you feared, brave Thor.'

'If you do not cease your mockery, Loki, I shall truly slay you.' Thor was now so angry that there is no doubt that he would have done as he threatened, had not Odin intervened. 'Peace, Thor,' he said, 'do not desecrate the halls of Aegir, who is our host.' And Loki rose and moved so that he stood between Thor and the door of the hall, while Thor lowered Miollnir and scowled furiously.

'Come, sit and drink with us, good Thor,' said Aegir, in his voice that was like the rumbling of the mighty waves. 'You have come late to the feast, but you are no less welcome for that. Tell us what adventures you have had. Did you slay many giants on your journey?'

'It is not a giant that I would slay at this very moment, if I had my way,' said Thor.

Loki stepped a little nearer to the door, for he knew of old Thor's thunderous rages, and feared lest there might soon be found no one, neither Odin nor another of the gods, to speak for him and so save him from Thor's wrath. He laughed again. 'I have said what I wanted to say to you all, and much mirth has it given me. But when Thor loses his temper, it is best to be far away.'

And all the gods cried out how they would be glad to see him go. 'And come not near us again, Loki, lest you regret it,' they called to him.

Loki shrugged his shoulders. 'I thank you for your good mead, Aegir,' he said, 'I shall not drink with you again. Farewell, Aegir, and farewell to you all, Aesir and Vanir. You have cast me out; seek not to find me fighting at your side when the end of all things comes.' And he was gone from the hall and away.

And there was no one there who was not glad to see him go, not even Sigyn, his wife. Yet Sigyn wept, while Frigg sought to comfort her; but she would not be comforted.

Loki, knowing that once he had declared himself their enemy, the hands of all the gods would be against him, hid himself on the top of a high mountain, in a house which had four doors, one in each of its four walls, so that he could see in all directions at one time and might not be taken by surprise. Close by the house there was a mountain stream and a waterfall, and here would Loki often swim in the shape of a salmon.

One day, as he sat in his house by the fire, to pass away the time he took twine and knotted it into a net, such as Ran, Aegir's wife, was wont to catch the sailors in. But suddenly he looked down into the valley and saw how Odin and Thor and others of the gods were climbing up the mountain side. 'Odin will have spied me from Hlidskialf,' he thought. 'But they will not find me here when they come.' And throwing his net into the fire, he went out from the house and leapt into the stream, changing himself into a salmon.

147

When the gods reached the house, they found it empty. 'He has gone,' said Thor angrily.

But Odin picked out from the ashes of the fire a half-burned net. 'He has been here but a short while ago,' he said. 'See, he has made himself a net for fishing.' And the gods looked out to where the stream flowed, and wondered if Loki had hidden himself in the water.

So they too made a net, and going to the stream, cast it into the water, and with Thor holding one side, and all the remaining gods the other, they dragged it up the stream. But Loki lay still between two stones on the bed of the stream, and the net went over him. A second time they cast the net into the water, having weighted it so that nothing might remain beneath it, and this time Loki leapt high out of the stream and the net passed under him.

The gods all called out to one another that they had seen him. 'He will try to swim down to the sea where he will be safe,' they said. And they went back again and divided into two groups, one on either side of the river, and so dragged the net along, down towards the sea, while Thor waded out into the middle of the stream, between Loki and the river mouth, and watched and waited.

And Loki saw that there were but two things left for him to do, to attempt to swim past Thor, down to the sea, or to jump once more over the net and swim back up the stream. And having little hope of evading Thor, he jumped again; but he was weary and he jumped not far enough, and Thor caught him by the tail; which is why, to this day, the salmon has a tail that is thinner than that of any other fish.

Loki struggled, but when he saw how all escape was hopeless, he took his own shape once again, that he might face whatever judgement the other gods put on him.

They took him to a cavern beneath the earth, and there

they bound him to a rock, so that he should remain there until the end of all things came. And Skadi, in her spite, now that Loki was powerless against her, brought a venomous serpent and fastened it above him, that the poison which trickled through its fangs might drip upon his face. 'So shall my father be avenged at last,' she said, and laughed.

But when all the gods were gone and had left him to his torments, Loki saw Sigyn come forward from the shadows. 'Why do you not go with the others?' he asked her.

She smiled through her tears and stroked his red hair. 'I would rather be with you,' she said. And she remained with him in the cavern, holding a cup above him to catch the drops of venom, so that it was only when she turned away to empty the cup that any of the poison touched him, and then he struggled so greatly that the whole earth trembled, and men called it an earthquake.

XX

The End of All Things

BUT the Norsemen believed that, as Odin had foreseen, the gods were doomed one day to perish, and this is how they told that it would come to pass.

First would there be three winters more terrible than any that had ever gone before, with snow and ice and biting winds and no power in the sun; and no summers to divide this cruel season and make it bearable, but only one long winter-time with never a respite. And at the end of that winter, Skoll, the wolf who had ever pursued the sun, would leap upon it and devour it, and likewise would Hati with the moon. And the stars which had been sparks from Muspellheim would flicker and go out, so that there would be darkness in the world.

150

The mountains would shake and tremble, and the rocks would be torn from the earth; and the sea would wash over the fields and the forests as Iormungand, the Midgard-Serpent, raised himself out of the water to advance on the land. And at that moment all chains would be sundered and all prisoners released; Fenris-Wolf would break free from Gleipnir, and Loki rise up from his prison under the ground. Out of fiery Muspellheim would come Surt the giant with his flaming sword; and out of her house would come Hel, with Garm the hound at her side, to join with her father, Loki. And all the frost- and storm-giants would gather together to follow them.

From Bifrost, Heimdall, with his sharp eyes, would see them come, and know that the moment which the gods had feared was at hand, and he would blow his horn to summon them to defend the universe. Then the Aesir and the Vanir would put on their armour, and the spirits of the dead warriors that were feasting in Valhall take up their swords, and with Odin at their head in his golden helmet, ride forth to give battle to the enemies of good.

And in the mighty conflict which would follow, all the earth, all Asgard, even Niflheim itself, would shake with the clang and cry of war. Odin would fight against huge Fenris-Wolf, and hard would be the struggle they would have. Thor, with Miollnir, would kill the Midgard-Serpent, as had ever been his wish to do; but he would not long survive his victory, for he would fall dead from the dying monster's poisonous breath.

Tyr and Hel's hound, Garm, would rush at each other and close to fight, and with his good left hand, brave Tyr would hew down the mighty beast; but in its last struggles it would tear the god to pieces, and so would they perish both.

Surt with his flaming sword would bear down on Frey,

but Frey had given his own sword to Skirnir, and as Loki had foretold, bitterly would he regret it, for he would have no more than the antler of a deer with which to defend himself. Yet would he not perish without a struggle.

As they had met and fought once before, over Freyia's necklace, so Loki and Heimdall would come together in battle once again, and Loki would laugh as he strove with his one-time friend. And in the same moment, each would strike the other a deadly blow, and both alike fall dead.

Though Odin would fight long and bravely with Fenris-Wolf, in the end that mighty monster would be too strong for him, and the wolf with his gaping jaws would devour the father of the gods, and then perish at the hands of Vidar the silent.

Then fire from Muspellheim would sweep over all, and thus would everything be destroyed; and it would indeed be the end of all things.

But the Norsemen believed that one day, out of the sea that had engulfed it, and out of the ruins, the world would grow again, fresh and green and beautiful; with fair people dwelling on it, born from Lifthrasir and Lif, the only man and woman to escape the fire. And they believed that out of the ashes of old Asgard would arise another home for the gods, where would live in joy and peace the younger gods, who had not perished; the two sons of Odin, Vidar the silent god, and Vali the son of Rind. And with them would be Magni, the strong son of Thor, mightier even than his father; while out from the house of Hel, at last, would come Balder, and Höd, his brother. And everywhere would be happiness.

An Alphabetical List of Names
Mentioned in the Stories

AEGIR: the god of the deep ocean.
AESIR: the gods who cared for mankind.
AGNAR: brother of Geirröd; protected by Frigg.
ALFHEIM: the home of the light elves, between earth and sky.
ALLFATHER: Odin, the king of the gods.
ANGRBODA: Loki's giant-wife.
ASGARD: the home of the gods.
ASK: the first man.
AUDUMLA: the cow whose milk was drunk by the giant Ymir, in the very beginning of things.

BALDER: the sun-god, son of Odin and Frigg.
BARREY: the forest where Gerd met Frey.
BAUGI: a giant, son of Gilling and brother of Suttung.
BERGELMIR: a giant; with his wife he was the only survivor when the other giants were drowned in Ymir's blood.
BESTLA: a giant-woman, Odin's mother.
BIFROST: the rainbow bridge between Asgard and Midgard.
BLODUGHOFI: Frey's horse.
BOLVERK: the name taken by Odin when he worked for the giant Baugi.
BORR: Odin's father.
BRAGI: the god of poetry, son of Odin and Gunnlod.
BRISINGA-MEN: Freyia's necklace.
BROKK: a dwarf who made a wager with Loki on the skill of his brother Sindri.
BRUNNAK: the home of Idunn and Bragi.
BURI: the first god.

DRAUPNIR: Odin's arm-ring, made by the dwarf Sindri.
DROMI: the second fetter with which the gods bound Fenris-Wolf.

EIR: the goddess of healing.
ELDIR: Aegir's serving-man.
ELLI: old age, the lord of Utgard's nurse.
EMBLA: the first woman.

FENRIS-WOLF: a huge wolf, son of Loki and Angrboda.
FENSALIR: Frigg's palace in Asgard.
FIALAR: a dwarf; together with Galar he killed Kvasir.
FOLKVANGAR: the place where Sessrumnir, Freyia's palace, stood.

FREY: the god of nature who ruled over the light elves; son of Niord.
FREYIA: the goddess of love and beauty; daughter of Niord.
FRIGG: the chief of the goddesses; Odin's queen.
FULLA: Frigg's handmaiden.

GALAR: a dwarf; together with Fialar he killed Kvasir.
GARM: the hound of Hel, the queen of the dead.
GEIRRÖD: brother of Agnar; protected by Odin.
GERD: a giant-maiden, beloved by Frey.
GILLING: a giant, father of Suttung and Baugi.
GINNUNGAGAP: the vast chasm in the beginning of things.
GIOLL: the river into Niflheim.
GLEIPNIR: the chain forged by the dwarfs with which Fenris-Wolf was bound.
GNA: Frigg's messenger.
GRIMNIR: the name taken by Odin when he went to King Geirröd's house.
GULLFAXI: the horse of the giant Hrungnir.
GULLINBURSTI: Frey's golden boar.
GUNGNIR: Odin's spear.
GUNNLOD: daughter of the giant Suttung; mother of Bragi.
GYMIR: a giant, father of Gerd.

HATI: the wolf who pursued the moon.
HEIDRUN: the goat who gave mead for the gods to drink.
HEIMDALL: the watchman of the gods.
HEL: the queen of the dead; daughter of Loki and Angrboda.
HERMOD: the messenger of the gods.
HLIDSKIALF: Odin's throne and watch-tower in Asgard.
HLIN: a goddess who served Frigg.
HÖD: the blind god of darkness; son of Odin and Frigg; twin brother and slayer of Balder.
HOFVARPNIR: Gna's horse.
HÖNIR: the bright god; Odin's brother.
HRAUDUNG: a king, father of Geirröd and Agnar.
HRINGHORNI: Balder's ship.
HRUNGNIR: the strongest of the giants; slain by Thor.
HUGI: thought, a youth in the lord of Utgard's house.
HVERGELMIR: the sacred spring that flowed in Niflheim by the root of Yggdrasill.
HYMIR: a giant, owner of a large brewing-cauldron.
HYRROKKIN: the giant-woman who launched Balder's funeral ship.

IDUNN: daughter of Ivaldi the dwarf; wife of Bragi.
IORMUNGAND: the Midgard-Serpent; child of Loki and Angrboda.
IOTUNHEIM: the home of the giant-people.
IVALDI: a dwarf, father of Idunn.

KVASIR: a wise man; killed by Fialar and Galar.

LAEDING: the first fetter with which the gods bound Fenris-Wolf.
LIF: the only woman to be saved alive at the end of all things.
LIFTHRASIR: the only man to be saved alive at the end of all things; from him and Lif a new race was to be born.
LOGI: wild-fire, a servant of the lord of Utgard.
LOKI: the god of fire.

MAGNI: Thor's son.
MIDGARD: the world of men.
MIDGARD-SERPENT: Iormungand, monster-child of Loki and Angrboda.
MIMIR: the giant who guarded the spring that flowed in Midgard by the root of Yggdrasill.
MIOLLNIR: Thor's battle-hammer.
MODGUD: the maiden who guarded Gioll, the river into Niflheim.
MUSPELLHEIM: the land of flaming fire.

NANNA: a goddess, Balder's wife.
NIFLHEIM: the land of mist where Hel ruled over the spirits of the dead.
NIORD: the king of the Vanir; god of the shore and shallow sea.
NOATUN: Niord's palace.
NORNS: the three fates: Urd, Verdandi, and Skuld.
NORSEMEN: the people who came originally from the countries now called Scandinavia, and who later settled in many parts of Europe.
NORTHLANDS: in this book, the lands now called Norway and Sweden.

OD: a god, Freyia's husband.
ODIN: the Allfather, king of the gods.

RAN: the goddess of the deep sea; wife of Aegir.
RIND: the mother of Vali.

SAEHRIMNIR: a boar in Asgard; slain and eaten each day, it came to life again each night.
SESSRUMNIR: Freyia's palace.
SIF: a goddess, Thor's wife.
SIGYN: a goddess, Loki's wife.
SINDRI: a dwarf, a skilled smith; brother of Brokk.
SKADI: daughter of Thiazi the storm-giant; wife of Niord.
SKIDBLADNIR: Frey's ship which could be folded up and carried in a wallet.
SKIRNIR: Frey's servant and friend.
SKOLL: the wolf who pursued the sun.
SKRYMIR: the name taken by the giant lord of Utgard when he met Thor and Loki on their travels.
SKULD: the future; one of the three Norns.
SLEIPNIR: Odin's eight-legged horse.
SURT: the giant from Muspellheim with the flaming sword.
SUTTUNG: a giant, father of Gunnlod.

SVADILFARI: the black stallion owned by the giant stone-mason who built the gods' citadel.
SVARTALFHEIM: the home of the dwarfs below the earth.
SYN: Frigg's doorkeeper in Fensalir.

THIALFI: Thor's servant.
THIAZI: the storm-giant; father of Skadi.
THÖKK: a giant-woman who would not weep for Balder; thought to be Loki in giant shape.
THOR: the god of thunder.
THRUDGELMIR: a six-headed giant.
THRUDHEIM: Thor's home in Asgard.
THRYM: the king of the frost-giants who stole Thor's hammer.
THRYMHEIM: the home of Thiazi the storm-giant.
TYR: the god of war.

URD: the past; one of the three Norns.
URD: the spring, guarded by the Norns, that flowed in Asgard by the root of Yggdrasill.
UTGARD: a fortress in Iotunheim.

VALHALL: the hall of the slain where the dead warriors feasted in Asgard.
VALI: son of Odin and Rind; he avenged Balder's death.
VALKYRS: Odin's warrior-maidens who carried the slain to Valhall.
VANIR: the gods of nature; Niord was their king.
VERDANDI: the present; one of the three Norns.
VIDAR: the silent god; Odin's son.

YGGDRASILL: the great ash-tree of the universe.
YMIR: the first giant.

A Note on Pronunciation

No one is absolutely certain of the pronunciation of the Old Norse language, but the following rules give an approximate pronunciation which should be close enough to the original for the average reader.

VOWELS

a (short): as in 'hat'.
a (long): as in 'father'.
e (short): as in 'men'.
e (long): as *a* in 'fate'.
i (short): as in 'is'.
i (long): as in 'machine'.
o (short): as in 'on'.
o (long): as in 'old'.
ö: as *ö* or *oe* in German 'schön' and 'Goethe'; the nearest equivalent sound in English is *er* in 'her'.
u (short): as in 'put'.
u (long): as *oo* in 'droop'.
y (short): as *i* in 'is'.
y (long): as *i* in 'machine'.

As a very rough indication of the length of a vowel, it may be taken that vowels are short (marked ˘ in example) when they are followed by a doubled consonant (e.g. Fŭlla) or two different consonants (e.g. Rĭnd); and long (marked ¯ in example) when followed by a single consonant (e.g. Vāli), although there are very many exceptions to this.

DIPHTHONGS

ae: as in Scots 'brae' or as *ay* in English 'day'.
ai: as *i* in 'fine'.
au: as *ou* in 'out'.
ei
ey }: as *ey* in 'they'.

157

The consonants may be pronounced as in English, but the following points should be noted.

G is always hard as in 'get', never soft as in 'gem'. When G comes before N at the beginning of a word (e.g. Gna) both letters should be sounded.

H before another consonant at the beginning of a word (e.g. Hrungnir) may be simply indicated by an initial breathing.

Y is used as if it were a vowel. See the note above.